Design Thinking

This book aims to provide readers with an in-depth understanding of design thinking by documenting the personal insights of professionals and practitioners from a wide range of disciplines.

Design Thinking: Theory and Practice refers to a series of cognitive, strategic, and practical steps used during the process of designing, and the context of how people reason when they engage with solving problems. The scope of this book focuses on topics such as problem-solving, systems thinking, innovation, and the role of design in product design and services. This book is unique as it brings together "stories" from both academics' and practitioners' perspectives, enabling readers to view design thinking from many different perspectives that can be applied in every-day life situations or for organizations when developing plans and policies.

This book would be essential reading for design engineers, industrial designers, and mechanical engineers who have interest in design thinking.

Design Thinking
Theory and Practice

Edited by
Eujin Pei and Kurt Becker

CRC Press
Taylor & Francis Group
Boca Raton London New York

CRC Press is an imprint of the
Taylor & Francis Group, an **Informa** business

Designed cover image: Adobe Stock ©

First edition published 2026
by CRC Press
2385 NW Executive Center Drive, Suite 320, Boca Raton FL 33431

and by CRC Press
4 Park Square, Milton Park, Abingdon, Oxon, OX14 4RN

CRC Press is an imprint of Taylor & Francis Group, LLC

ISBN: 978-1-032-76780-2 (hbk)
ISBN: 978-1-032-78363-5 (pbk)
ISBN: 978-1-003-48752-4 (ebk)

DOI: 10.1201/9781003487524

Typeset in Times
by KnowledgeWorks Global Ltd.

Dedication

To innovators and changemakers who dare to reimagine the world, and to design educators who inspire us to think differently. This book is dedicated to our wives, Ying and Mariam, whose dreams we share and whose unwavering support that has made this journey of writing possible.

Contents

Foreword

John Gero
University of North Carolina at Charlotte

Designing is a foundational activity of humans. It is the way humans intentionally change the world around them with the goal of making it better for their purposes. Until the 20th century, designing was restricted to large-scale products such as machines, ships, civil structures, and buildings. However, the development of consumer goods and then software applications has changed the landscape of design and noticeably broadened the scope of design. Designing has moved beyond its historical roles to encompass the design of organizations, and social and governmental systems. The range of design has spawned the creative industries, a catch-all phrase that encompasses designing well beyond its traditional areas. The creative industries contribute around $1 trillion a year to the global economy. Designing is one of the foundations of the creative industries. Designing has become a significant activity in both the national and global economy. As a consequence, it plays an important role in economic and social wealth generation. What is surprising is that given the ubiquity of designs and its continuous history from the earliest days of civilization (the *Epic of Gilgamesh*, written around 4,000 years ago, mentions designing), how recently designing has become an area of research. Designing had been learned through the master-apprentice model for generations. Formal research into designing began in earnest only post World War II, in the 1950s, although there were earlier individual efforts. This research has as its goal to produce a body of knowledge that would lead to the understanding of designing as a set of activities that could be formalized, taught, and potentially improved. The teaching required that there be knowledge to teach.

The knowledge of designing could be obtained through induction from observations of designers in action. Or this knowledge could be generated through executing experiments that elucidated different aspects of designing. These two streams of knowledge generation intertwine with each feeding off the other. Experiments produced models of designing based on either behaviors of designers or concepts from cognitive science, i.e., how the mind behaves. Both of them have been labeled "design thinking" and this has caused confusion as to which is being referred to. So, design thinking can be what designers do or it can be what happens in the minds of designers, which results in what designers do. The reader of this volume will find examples of both streams.

What both these streams agree upon is that designing is a set of processes, where the processes can be individually delineated. The confluence of these processes produces designing and it is this set of processes that distinguishes designing from other human activities.

This current design thinking model of designing, derived from design practice, distinguishes itself from earlier models through the inclusion of the user in the process under the label of "empathy". This notion of empathy, imagining the world from multiple perspectives, was missing as a significant factor from earlier models of design thinking, such as Asimov's 1962 model of engineering design (Asimov, 1962) and in Pahl and Beitz's 1984 model of engineering design (Pahl & Beitz, 1984). The highly influential book by Herb Simon, *The Sciences of the Artificial* (Simon, 1969) also does not address this notion of empathy in its description of designing. It was an article in the *Harvard Business Review* by Tim Brown, CEO of IDEO (Brown, 2008), that initiated the widespread interest in design thinking. The article was expanded into a highly influential book (Brown, 2009). Brown distilled what he had learned from his own design practice to produce a near-prescriptive model of design thinking and framed it within a business perspective that elevated designing from an activity associated with producing products to one generating value for an organization.

The other stream, exemplified by Lawson (Lawson, 2005) and Cross (Cross 2023), focuses on the cognition of designing within a process model. The two streams complement each other.

This volume brings together 12 contributions on design thinking that span both streams and contribute to the literature in unique ways. In the introductory chapter, Becker outlines the foundational activities that go to make up design thinking. In Chapter 2, Cabrera focuses on the way design thinking contributes to innovation within the social context. In Chapter 3, Huck and Sukhov dive more deeply into the role of ideas and their sharing in design development. In Chapter 4, Mesa and Ruiz-Paster, in contrast to Chapter 3, expand the scope of designing to cover the circular economy within a design thinking framework. In Chapter 5, Morcos brings the concept of the empathy aspect of design thinking to the start-up culture. In Chapter 6, Mortati, Mariani and Rizzo bring design thinking to bear when developing public services, exemplified by projects in the European Union. In Chapter 7, Kelly and Grace pursue design thinking from a more cognitive aspect when considering a framework for the role of human designers in an AI-human hybrid intelligence. In Chapter 8, Genert, Stenftenagel and Blandon take a more traditional systems-subsystems view of design thinking as they develop a Multidisciplinary Design and Optimization model of their designing activity. This model generally excludes direct human aspects. In Chapter 9, Kannengiesser and Gero look at the empirical data that supports the human cognitive behavior of the dual process modes of thinking when designing. In Chapter 10, Hahri, Lubis and Pei look at Co-Design from a design thinking perspective. In Chapter 11, Palmås pursues an example of one of the processes that makes up design thinking, namely, the creation of ideas using metaphors and its place in design education. In Chapter 12, White, Patocs, Beauchamp and Raina, fittingly as the final chapter of this volume, present results from a large scale seven-year study of using design thinking.

This volume brings material not normally found together in a single place. It brings design thinking from a cognitive perspective along with design thinking from the empathy perspective. This shows both the breadth and depth of the two streams of design thinking and how they complement each other. Recent research on the brains of designers as they design has shown that both streams provide frameworks for what happens in the brain (Gero & Milovanovic, 2023).

REFERENCES

Asimov, M. (1962). *Introduction to Design*. Prentice-Hall.

Brown, T. (2008). *Design Thinking*. Harvard Business Review.

Brown, T. (2009). *Change by Design: How Design Thinking Transforms Organizations and Inspires Innovation*. HarperBusiness.

Cross, N. (2023). *Design Thinking: Understanding How Designers Think and Work*, 2nd Edn. Bloomsbury VA.

Gero, J. S. & Milovanovic, J. (2023). Design thinking and design neurocognition, in K. Straker & C. Wrigley (eds), *Research Handbook on Design Thinking*. Edward Elgar, pp. 7–24. https://doi.org/10.4337/9781802203134

Lawson, B. (2005). *How Designers Think*, 4th Edn. Routledge.

Pahl, G. & Beitz, W. (1984). *Engineering Design: A Systematic Approach*. Springer.

Simon, H. A. (1969). *Sciences of the Artificial*. MIT Press.

Preface

Design Thinking: Theory and Practice is more than a guide. It is a call to action that invites designers, engineers, policy-makers, educators, and innovators to rethink their approaches and embrace a mindset that prioritizes adaptability, empathy, and collaboration. To be successful in today's highly technological and globally competitive world, one requires the use of a different set of skills. This book seeks to empower readers to leverage design thinking not only as a tool for innovation but also as a means to create a better and more sustainable world. Through case studies, theoretical frameworks, and empirical research, it challenges readers to consider whether and how design thinking can sustain long-term relevance in evolving contexts. As the chapters unfold, readers are invited to journey through the diverse dimensions of design thinking. From understanding its foundational elements to witnessing its integration with modern disciplines such as using artificial intelligence, product design, and circular economy principles, the narrative of this book highlights the versatility of this approach. The book offers insights into how design thinking can drive change across industries and domains. In essence, this book presents a comprehensive exploration of design thinking principles, methodologies, and applications, extending from the design of everyday products to strategies for addressing global challenges.

About the Editors

Eujin Pei began his career as a product designer, with his work exhibited at the Cooper Hewitt Smithsonian Design Museum in New York. He now works as a Full Professor and serves as Associate Dean for the College of Engineering, Design and Physical Sciences at Brunel University of London, UK. Since 2015, Eujin has been the Director of the BSc Product Design Engineering course, where he leads its development and ensures graduates are equipped with the skills to excel in a rapidly evolving industry. Before this, he served as Director for Postgraduate Research, where he played a key role in maintaining academic standards and promoting excellence among PhD students. Eujin is a Fellow of the Institution of Engineering Designers (FIED), and his professional qualifications include being a Chartered Engineer (CEng), Chartered Environmentalist (CEnv), and Chartered Technological Product Designer (CTPD). His work reflects a strong commitment to innovation, sustainability, and the advancement of technology in design and engineering.

Kurt Becker spent his career as an engineering and technology educator and has extensive international experience working on engineering education and technical training projects funded by the Asian Development Bank, World Bank, and U.S. Agency for International Development (USAID). Countries where he has worked include Armenia, Bangladesh, Bulgaria, China, Egypt, Indonesia, Macedonia, Poland, Romania, and Thailand. Kurt is Professor Emeritus in the College of Engineering at Utah State University (USU) where he spent 30 years working in the discipline of Engineering Education. He was Department Head and Founder of the Department of Engineering Education at USU. His research in engineering education spans more than three decades, and his work preparing PhD students to be the next generation of leaders and researchers demonstrates his commitment to the discipline. While in the College of Engineering at USU he stated the National Academy of Engineering endorsed, "Grand Challenges Scholars Program". This program provides undergraduate engineering students with curricular and co-curricular activities while being mentored by engineering faculty. Kurt's work in engineering education and design thinking is a tribute to his dedication to the field of engineering and technology.

List of Contributors

Marla Beauchamp
McMaster University, Canada

Kurt Becker
Utah State University, USA

Nick Blanton
Lockheed Martin Missiles and Fire
Control, USA

Sixto Cabrera
Hardware Engineer, USA

Jaime Albero Mesa Cogollo
Universidad del Norte, Colombia

Glenn Gebert
Lockheed Martin Missiles and Fire
Control, USA

John S. Gero
University of North Carolina at
Charlotte, United States

Kazjon Grace
The University of Sydney, Australia

Jana Huck
Karlstad University, Sweden

Udo Kannengicsser
Johannes Kepler University, Austria

Nick Kelly
Queensland University of Technology,
Australia

Pierre Lubis
University of Waikato, New Zealand

Ilaria Mariani
Politecnico di Milano, Italy

Karim Morcos
Imperial Studio, Egypt

Marzia Mortati
Politecnico di Milano, Italy

Karl Palmås
Chalmers University of Technology,
Sweden

Laura Ruiz-Pastor
Universitat Jaume I, Spain

Audrey Patocs
McMaster University, Canada

Eujin Pei
Brunel University of London, UK

Parminder Raina
McMaster University, Canada

Francesca Rizzo
Politecnico di Milano, Italy

Eric Stenftenagel
Lockheed Martin Missiles and Fire
Control, USA

Alexandre Sukhov
Karlstad University, Sweden

Bahareh Shahri
University of Canterbury, New
Zealand

P.J. White
South East Technological University,
Ireland

Acknowledgments

This book is the brainchild of a chance meeting between two engineering educators from different continents who care deeply about engineering and how design thinking is critical to the engineering discipline. This book would not have been possible without the invaluable contributions of authors and co-authors for sharing their knowledge so generously, and whose expertise and dedication have enriched its pages. A special thanks to Shatakshi Singh and Katya Seward from Taylor & Francis for their unwavering guidance and support in bringing this project to fruition. Thanks to our friends and family members also who have supported us along the way. This book is a testament to the collaborative spirit that underpins the very essence of design thinking.

1 Introduction to Design Thinking

Kurt Becker

Today, organizations pursue innovation to be successful. Innovation can increase the bottom line, improve quality, or make processes more efficient and productive (Magistretti et al., 2021). Apple Inc. is a good example, as it has had a core value of innovation dating back to its formation, and the result is an array of products setting it apart from its competitors. Innovative goods and services are now synonymous with superior quality. The more an individual or organization demonstrates a fuller understanding of innovation, the greater is the propensity to attain innovation (Kahn, 2018).

The one constant thing in today's work environment is change, and the rate of change is increasing. The global economy drives change and has become the norm today. Companies and institutions that fail to adapt to change will lose their competitive edge (Igbal, 2011; Çetinkaya et al., 2019). An example includes Research in Motion Limited, the original company of BlackBerry Limited that was once a leader in wireless data technology, but it failed to innovate and keep up with the change in the industry. They are now struggling to regain market share. This is just one of many examples of companies that have failed due to lack of innovation. While innovation is often thought to start the design process, the innovative solution is discovered towards the end of the design process. This is because using the design thinking process requires the user to think deeply about the design solution and that the context is well understood.

1.1 DESIGN THINKING

To complement innovation, the concept of design thinking has gained momentum in recent years. In many organizations, it has become a crucial element in the strategic planning and decision-making processes. Design thinking is a methodology for innovation focused on understanding and solving complex customer problems. It emphasizes the importance of empathy for the user, a hands-on approach to problem-solving, and the iteration of solutions that address specific user needs (Nakata, 2020). According to Brown (2008), design thinking has emerged as a human-centred approach to innovation based on the ways that designers think and work. By incorporating design thinking into their business strategies, companies can gain a competitive edge by creating products and services that truly resonate with their target audience (Nakata & Hwang, 2020; Jaskyte & Liedtka, 2022). Design thinking is a practical, creative problem-solving method that leads to solutions with the intent of an improved future result (Foster, 2021).

Design thinking is a method often associated with creativity and developing new ideas. It is used in problem-solving and development of solutions for complex issues. In early use, design thinking was used in the development of specifications for a product or service to define how the product or service would be used by the customer or client. Over time, it has evolved and become a new discipline (Brown, 2009).

Design thinking is a process that involves using many different skill sets and approaches, and is not just about creating new products or services. It is about defining new ideas, creating new ways of doing things, and implementing innovative solutions to complex problems (Auernhammer and Roth, 2021). Design thinking is a multidisciplinary approach that draws on the skills and knowledge of experts from a range of fields. Working together in a collaborative way, these experts identify underlying patterns of problems and create new and innovative solutions to those problems (Verganti et al., 2021).

Tim Brown, CEO of IDEO, defines design thinking as a methodology to understand user demands and requirements, challenge the assumptions that define issues that are not easily understood, and create innovative solutions by considering conditions and resources. According to Dell'Era et al. (2020), design thinking is accepted as a formal creative problem-solving method with the intent to foster innovation. This description has been based on how practicing designers claim to approach problems, rather than on how design has been defined in academic literature (Brown 2008; Pande & Bharathi, 2020; Cross, 2023).

At its core, design thinking is a problem-solving approach that is analytical and creative and combines logic, intuition, and systemic reasoning to explore the possibilities of what could be and create desired outcomes that benefit the end user (i.e. the customer). According to Kim & Park (2021), a key aspect of design thinking is that it is human-centred; it cares about the people who will utilize what is created and seeks to understand their wants, needs, and experiences. This way, it is possible to understand the extent of the problem and create a deep understanding of the people the solution aims to help. Design thinking also allows for greater collaboration between teams increasing the chance of creating a successful product or service. This is because, from the outset, it creates a culture that welcomes the sharing of ideas, and it decreases risk as the proposed solutions are well thought out. Figure 1.1 shows designers using the design thinking process.

1.2   ORIGINS AND KEY CONTRIBUTORS

Design thinking has evolved since the 1960s, drawing from diverse influences and approaches, and has paralleled this conceptual development as a framework of methodologies supporting complex problem thinking (Henriksen, et al., 2017). According to Jamal et al. (2021), "design thinking is a human-centric, structured,

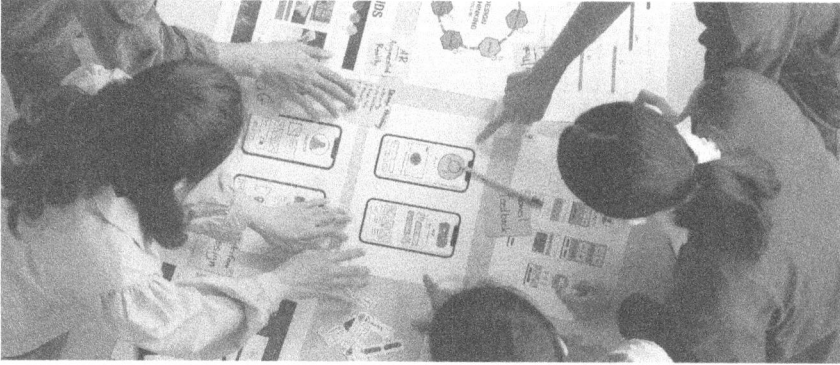

FIGURE 1.1 Designers using design thinking. (Google image – purchased) – web site https://stock.adobe.com/plans.

collaborative problem-solving approach that produces innovative products, processes, or experience solutions to address wicked problems" (p. 3).

The design thinking process practiced today was created through the combined efforts of many prominent engineers, economists, and designers over the span of decades. Designers such as Herbert Simon, L.J.K. Smith, and John Chris Jones are identified as early design thinkers, having theorized about the processes by which designers go about their work (Thawani, 2023). According to Thawani, they did this intending to describe these processes in such a way that non-designers could understand and potentially apply them to their professional work contexts. Discussions about design thinking processes and practices were fomented by early adopters like Richard Buchanan, Nigel Cross, and Tom Inns. More modern discussions and applications of design thinking have increased exponentially in recent years, and more high-profile stakeholders in contemporary design thinking include Tim Brown of IDEO, Roger Martin of the Rotman School, and David Kelley of the Stanford d.School (Muratovski, 2021).

1.3 DESIGN THINKING PROCESS

Design thinking utilizes elements like empathy and experimentation to arrive at innovative solutions. This process often involves much collaborative work in which interdisciplinary teams work together to solve problems and arrive at solutions. The design thinking process is best thought of as a system of overlapping spaces rather than a sequence of orderly steps. More specifically, the design thinking process is a method of generating useful and creative ideas and is also outlined as a cognitive, strategic, and practical process by which design concepts are engineered. There are several variations of the design thinking process in use today, and they have three to seven phases. However, all variations of the design thinking process are similar and consist of three main phases: Define, Design, and Test (Brown, 2008; Knight et al., 2020).

Double Diamond Model

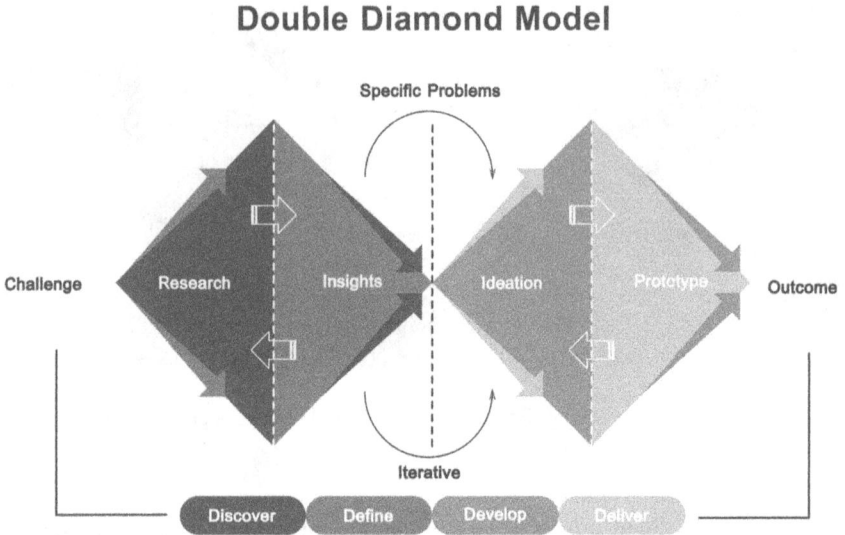

FIGURE 1.2   Double diamond design thinking model. (Google stock photo)

A commonly used version of design thinking consists of six chronological steps: *empathize, define, ideate, prototype, test,* and *implement.* Designers looking for solutions to a problem, using this process, first understand it better (empathize) and then use the understanding to get a clearer problem statement (define). The outcome is a never-ending process of problem-solving through creating and evaluating.

Today, many different models and schools of thought exist for design thinking. Whether it is d.school at Stanford University, IDEO U, MIT Sloan, or any of the many other schools, the design thinking process consists of multiple phases. These range from the four-phase double diamond model (see Figure 1.2) to the five-phase design thinking model (see Figure 1.3), to the six-phase design thinking model (see Figure 1.4).

Although there is no single model of the design thinking process, for this chapter, it can be explained by a five-phase model. The following section discusses each phase of the design thinking process.

1.3.1   Phase 1. Empathize: Understanding User Needs

The first phase of the design thinking process is to empathize. During this first step, the designer should empathize with the person who will use the product. This involves observation, engaging with, and learning from the person in his or her environment. This phase starts with observing and interviewing the end user about the problem or product needs. This phase constructs a situation map of the user's experience. After understanding the real problem or project needs, get an agreement with the user to make sure this is a real problem. This enables the designer to identify the needs to solve the user's problem (de Kervenoael et al., 2020).

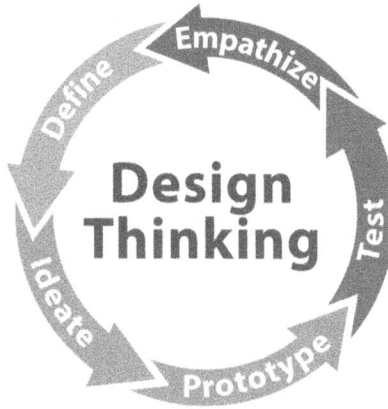

FIGURE 1.3 Five-phase process of design thinking.

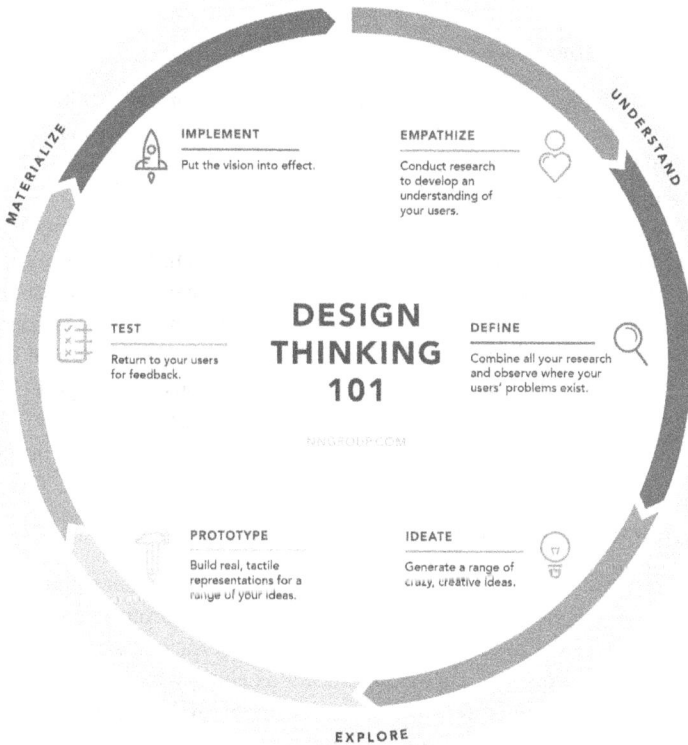

FIGURE 1.4 Six-phase process of design thinking. (Google stock photo)

1.3.2 PHASE 2. DEFINE: FRAMING THE PROBLEM

The problem with much of design today is that it is poorly understood, undefined, or confusing. A clear statement on what the problem is provides more information than numerous vague statements. The "define" stage is crucial to the rest of the process. This is where you, as a designer, can understand and articulate the problem. This may involve several sessions, with the design team present, to brainstorm on what has been learned in the empathize stage. Outputs of the sessions should be consolidated into a concise problem statement (Henriksen et al., 2017). Often this is displayed in a place visible to the design team and referred to continuously during the project to keep focus.

During the define phase, you should try to get a clear understanding of what is valuable and a clear understanding of the problem. What are you trying to achieve? It is quite easy to prematurely get the problem defined down into solutions, such as product specifications. This is a premature act and should be avoided. A solution or solutions come later. You will find that it is an extremely iterative process, where you will change your mind many times. Throughout, you should maintain primary focus on the people you are designing for, and the problem is defined by their needs. So, it is important to keep an open mind and seek feedback from those you are designing for.

1.3.3 PHASE 3. IDEATE: GENERATING CREATIVE SOLUTIONS

The idea phase combines all the background information and creates ideas of how to better solve the existing problem. The end goal of this phase is to develop potential solutions to the problem. In this phase, it is also important to observe constraints and keep an open mind (Waidelich et al., 2018).

The design team, with a clear understanding of the problem, should generate ideas and produce the optimum solution to the problem. Brainstorming is an often-used method, but there are many ideation techniques available to get the best output from the team. Examples of other ideation techniques include, reverse the problem, Rapid 8s (a fast-sketching exercise to get people to produce 8 ideas in a short amount of time), and "How might we" questions to help you produce creative ideas while focusing on the problems that you are trying to solve. A successful ideation session should provide many potential solutions to the problem.

1.3.4 PHASE 4. PROTOTYPING: BUILDING AND TESTING IDEAS

In this phase, it is important to narrow all the ideas in the previous phase to one. Once one idea is selected, a prototype is constructed. The prototype should be a simplified version of the final product or idea. A prototype can affect implementation by cheaply and quickly answering questions about what was implied by a requirement, resolving potential disputes or conflicts, and uncovering new and more efficient requirements. This will allow you to learn about the possibilities and constraints of your idea and further identify potential issues (Luka, 2014).

The prototype can be used for internal use by the designers and implementers to aid in the understanding of the existing system and bring about a new one. The prototype can also produce an object, a proposal, or a demonstration of a system that can be invaluable for evaluation considering new understanding by observing, manipulating, or viewing the prototype.

1.3.5 PHASE 5. TEST: GATHERING FEEDBACK AND ITERATING

During the testing phase, the development team should adhere to feedback from the users. This includes observing and taking note of what the user likes or dislikes. This feedback should be organized and given to the developers in a format that they can use to evolve the prototype into an improved solution. This is an iterative process, and high-quality solutions can be identified by many iterations and the feedback for refined changes (Mueller-Roterberg, 2018).

In design thinking, testing is the ultimate manifestation of the iterative nature of the process. This is where all the previous modes of thinking, especially the generation of empathic understanding, the development of insight, and the construction of well-defined problems pay off. Testing is the stage that sheds light on the solution and whether it successfully solves the problem.

1.4 DESIGN THINKING IN DIFFERENT INDUSTRIES

Design thinking has brought about significant transformation in various industry sectors and it is proving successful in solving complex problems. From simple products, like home kitchenware and furniture, electronics, and jewellery, to complex systems, like city management, educational systems, health care processes, and financial systems, are being designed, developed, and implemented using the concepts and principles of design thinking. A few industries using design thinking include business, technology, healthcare, and education (McLaughlin et al., 2019; Panke, 2019; Pande & Bharathi, 2020).

Design thinking is not limited to facilitating product innovation; it also can and should be applied to service innovation. Unfortunately, much service innovation involves only process innovation rather than new service offerings and can often degenerate into cost-cutting exercises. This is because applying traditional methods of innovation such as Six Sigma, lean production, or even certain types of marketing can lead to an increased focus on cost efficiency and internal problems rather than creating value for the customer (Gloppen, 2009). The strength of design thinking is that it keeps the user at the centre of all innovation efforts to determine what their needs are (Verganti et al., 2021). By understanding and observing user behaviour, it is possible to get to the root of problems and create solutions that users need and want. An example of this is the various, well-documented methods of IDEO. According to Ortiz (2016), when developing a new service for Prada, a prestigious brand in New York City, IDEO did not just look at what could be changed about the current service, but listened to the client demands and observed the behaviour of a client in the store and level of independence when trying clothes on.

Frequently, the experience of the service and the experience of the product are intertwined, and creating a comprehensive approach is integral to complete customer satisfaction (Bell 2008). More importantly, the nature of services is such that they are intangible and often unclear to both provider and consumer, leading to a great opportunity to create and clarify new value. Services are all around us and can be categorized into those that are personal (e.g. banking) and those that are public (e.g. health care). In the modern, competitive, global business climate, even public services must market themselves and gain user satisfaction. An example of this is the UK's National Healthcare System (NHS), which can now be evaluated with star ratings, just as one would rate a hotel (Chang, 2009).

Design has been acknowledged as an important educational skill for some time, and design thinking has become an approach to teaching and learning for 21st-century skills and knowledge (Becker & Mentzer 2015; Hennessey & Mueller, 2020). Design thinking promotes a creative problem-solving method fostering innovation (Dell'Era et al., 2020). Because current and future generations will be confronted with problems of greater complexity and higher stakes than those of the past, there is a compelling need for young people to be better problem-solvers and to be comfortable with the kinds of open-ended problems that characterize technological, social, and environmental challenges. By integrating design thinking into education, students might increase their ability to deal with ambiguity and improve their persistence when dealing with difficult but important problems (Hennessey & Mueller, 2020).

Design thinking education can take on many forms, from workshops to projects, and it has been implemented in schools ranging from elementary schools to universities (Hennessey & Mueller, 2020). Regardless of how design thinking is taught, the tools and methods remain valuable assets for students to learn and employ in their studies and careers (Noh & Karim, 2021).

Design thinking has a place in the curriculum when the skills and thought processes are considered in a systematic way (Rusmann & Ejsing-Duun, 2022). Design thinking skills are crucial for today's students as the world is filled with complex ever-changing open-ended challenges. Design activities aimed at bringing real-world problems to the classroom often help students experience the need for understanding, synthesizing, and utilizing information from several different resources and in various forms. For example, having students work with industry on actual problems will help students learn about the various steps of design thinking. This can help students with the creative thinking process and serve as a catalyst to motivate them to be more creative (Balakrishnan, 2022).

Although design thinking is a creative, dynamic, and systematic process for problem-solving, it is new to technology. It is becoming more popular due to competitive products continually being released. Problems occur frequently in the technology industry and design thinking is a perfect problem-solving method. It is a way for designers to understand and solve a problem. This is important in technology where satisfying user requirements is essential (Nakata & Hwang, 2020; Liu & Lu, 2020).

Healthcare is another industry that uses design thinking to solve problems. Some challenges modern-day healthcare systems face includes elderly care, increasingly more expensive treatments, and a lack of affordable preventative care. Globally, healthcare is a growing industry, with huge variations in how medicine is practiced. In recent years, many large healthcare institutions have started to employ designers to work on new product and service development (Kolko, 2015). Since health management is in demand, the importance of patient tracking and managing patient follow-up for chronic care has grown. However, creating efficient and inexpensive solutions to these problems is complex (Thakur et al., 2021).

An early example of design thinking being used in healthcare is the use of a home pregnancy test kit (see Figure 1.5). A freelance graphic designer working at a pharmaceutical company saw hundreds of pregnancy tests that doctors had sent from their offices in the company's lab. "It's so simple, just a test tube and a mirrored surface. A woman could do that herself" (Catlin 2015). Catlin added, after trial and error, she created in 1967 a prototype home pregnancy test (see Figure 1.5). Though the graphic designer was unaware at the time, she used the design thinking process to develop the prototype. She used "empathy" to understanding user needs, she "defined" by framing the problem, "ideated" to generate creative solutions, "prototyped" to building and test ideas, and she "tested" to gather feedback and iterate.

Crane's name was on the patents for the device, which the pharmaceutical company, Organon, licensed to companies that brought the early pregnancy test (e.p.t.), the Answer and Predictor to market in 1977 (Google Patents/USPTO).

This product allows the patient to self-diagnose in their home, saving time and resources for care providers. This allows the patient a more comfortable and less anxious environment to receive the test results. This practice is cheaper, easier, and requires less resources than the traditional method of pregnancy testing. Design thinking in healthcare can lead to less waste, improvements in preventative care, and patient self-management.

1.5 CHALLENGES OF DESIGN THINKING

The process of design thinking is generally very positive. It has emerged as an important way for designers to draw on rich customer insights to enhance their products and services (Knight et al., 2020). Since the design thinking process takes time and money to get started, main challenge is that it is difficult for organizations to justify short-term loss for long-term gain. This is increased since there is no guarantee of the outcome of design thinking. It is human nature to resist change, and design thinking is a change from the status quo. It takes people from where they are now to an improved state, and it can help organizations overcome cognitive challenges when transitioning to new innovation approaches and outcomes (Kolko, 2015; Liedtka, 2015; Randhawa et al., 2021; Cross, 2023).

Another challenge in design thinking is finding a balance between creativity and practicality. For design thinking to be truly effective, it should strike the right

May 18, 1971   M. M. CRANE   3,579,306

                                   DIAGNOSTIC TEST DEVICE

                                     Filed Jan. 22, 1969

FIG.I

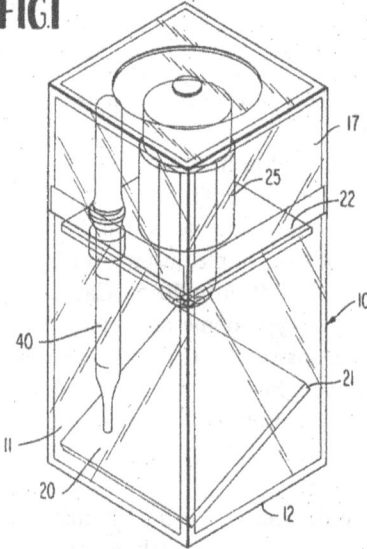

FIG.2

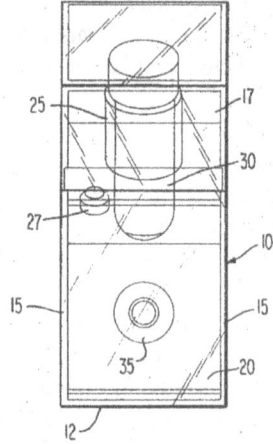

FIG.6

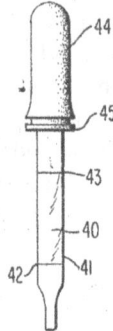

FIG.5

FIG.4

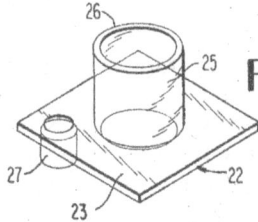

FIG.3

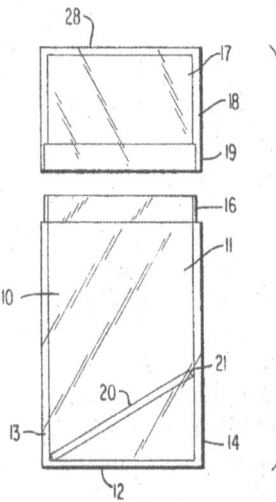

INVENTOR
MARGARET M. CRANE

BY   Hugo E. Weisberger
                                ATTORNEY

FIGURE 1.5   Patent drawings of a prototype home pregnancy test device.

balance between creativity and practicality (Auernhammer & Roth, 2021). The primary focus should be on creativity, but in some cases of design thinking "*creative problem solving* relies on combining the conscious and unconscious mind, rational thought, and imagination" (Dell'Era et al., 2020, p. 330).

In today's work environment, there is resistance to change (Malhotra et al., 2021). This can threaten innovation more than anything else. Design thinking is a disruptive process, and change is often difficult and messy. According to Conway et al. (2017), "While design thinking has proved itself to be successful in the realm of creating new products and services, the challenge is how to support innovations to enter and actively shape the complex systems that surround wicked social challenges" (p. 3). Empathetic and experimentation are qualities of design thinking that threaten those who do not feel comfortable taking chances, and the design thinking methodology clashes with the traditional strategic learning model. Schools and students ingrained with the conventional learning model are disadvantaged in employing design thinking. Long-time practitioners who use analytical and convergent thinking skills might be averse to switching to trial and error, divergence, and synthesis (Knight et al., 2020; Lynch et al., 2021).

1.6 CONCLUSION

Innovation is critical to the way the world advances. Often, people think about technology when they think about innovation, but it is much broader than that. Innovations are taking place daily, all around us, and the more an individual or organization demonstrates a fuller understanding of innovation, the greater is the propensity to attain innovation (Kahn, 2018). In this chapter, an introduction to design thinking has been discussed. Applications of design thinking in new areas, such as organizational change, leadership, and education have been discussed (Verganti et al., (2021). Design thinking is a novel problem-solving methodology well suited to the challenges organizations face in encouraging more creative thinking and achieving innovation and growth (Liedtka, 2018).

There is no denying that no matter what the industry, the impact of design thinking as an innovation is substantial. Users have increased demands and higher expectations than ever before and are looking for products and services to meet those demands. The importance of design thinking is growing because this approach can be implemented as a human-centred process that uses a step-by-step approach which prevents fluency in thinking and flexibility in approach, which are essential in design innovation (Auernhammer & Roth, 2021). Furthermore, as the elements of design thinking become recognized, more traditional thinkers can become more innovative in their approaches. Design thinking is not only for designers but also for creative employees and leaders who seek to incorporate design thinking into every level of an organization, product, or service (Dam & Siang, 2021). Thus, the evolution of design thinking to drive new alternatives for business and society has real-world, wide-spread implications (Lewis et al., 2020; Pande & Bharathi, 2020).

When exploring the design thinking process, five phases have been emphasized. During the empathy phase, an understanding of the people for whom you are designing is highlighted. This is not to be confused with market research. The define phase helps find opportunity areas for design. This phase helps transition from the understanding of the users' needs and problems to achieve specific goals. The ideate phase is the mode of idea generation. During ideation, idea generation should be recorded so all ideas are presented. During the prototype phase, preliminary solutions generated in earlier phases are developed into physical working models. This is the experimental phase of the process where you identify the best possible solution. The test phase is testing it out. The goal is to identify any problems with the solution and, if necessary, return to previous phases. The aim is to refine your solution. You will likely go around this cycle several times before coming to a final solution. The design thinking phases are a guide that help keep the design process on course and with continued emphasis on what is most important to the user.

Design thinking is a powerful method of problem-solving that begins with understanding customer needs. It is distinguished by its iterative nature, user orientation, and its concern for providing a better solution (Brown, 2008; Kim & Park, 2021). In the past 50 years, there has been a quiet revolution in the way design is understood and executed. This movement has been led by a new, multidisciplinary generation of designers who apply design thinking to the increasingly complex problems that contemporary society faces. Designers are using these methods to create new products or services to identify opportunities for businesses leading to a more desirable experience for customers.

The perceived benefits of design thinking are that it delivers innovation, improves products and services, and creates better customer value through a more product-focused development approach (Nakata & Hwang, 2020). As long as it keeps delivering results, the methodology will continue to evolve, and the design thinking process will help to enhance creativity and innovation based on designers' practices (Brown, 2008; Auernhammer & Roth, 2021).

REFERENCES

Auernhammer, J., & Roth, B. (2021). The origin and evolution of Stanford University's design thinking: From product design to design thinking in innovation management. *Journal of Product Innovation Management, 38*(6), 623–644.

Balakrishnan, B. (2022). Exploring the impact of design thinking tool among design undergraduates: A study on creative skills and motivation to think creatively. *International Journal of Technology and Design Education, 32*(3), 1799–1812.

Becker, K., & Mentzer, N. (2015, September). Engineering design thinking: High school students' performance and knowledge. In *2015 International Conference on Interactive Collaborative Learning (ICL)* (pp. 5–12). IEEE.

Bell, S. (2008). *Design thinking. Temple University Libraries.*

Brown, T. (2008). Design thinking. *Harvard Business Review, 86,* 84–92.

Brown, T. (2009). *Change by design: How design thinking transforms organisations and inspires innovation.* New York: HarperCollins.

Catlin, R. (2015). The unknown designer of the first home pregnancy test is finally getting her due. *Smithsonian Magazine.* Museums Correspondent, September 21, 2015.

Çetinkaya, A. Ş, Niavand A., & Rashid, M. (2019). Organisational change and competitive advantage: Business size matters. *Business and Management Studies: An International Journal, 7*(3), 40–67. http://dx.doi.org/10.15295/bmij.v7i3.1230.

Chang, L. C. (2009). The impact of political interests upon the formulation of performance measurements: The NHS star rating system. *Financial Accountability & Management, 25*(2), 145–165.

Conway, R., Masters, J., & Thorold, J. (2017). *From design thinking to systems change. How to invest in innovation for social impact. RSA Action and Research Centre.*

Cross, N. (2023). *Design thinking: Understanding how designers think and work.* Bloomsbury Publishing.

Dam, R. F., & Siang, T. Y. (2021). What is design thinking and why is it so popular? Design Foundation.

de Kervenoael, R., Hasan, R., Schwob, A., & Goh, E. (2020). Leveraging human-robot interaction in hospitality services: Incorporating the role of perceived value, empathy, and information sharing into visitors' intentions to use social robots. *Tourism Management, 78,* 104042.

Dell'Era, C., Magistretti, S., Cautela, C., Verganti, R., & Zurlo, F. (2020). Four kinds of design thinking: From ideating to making, engaging, and criticizing. *Creativity and Innovation Management, 29*(2), 324–344.

Foster, M. K. (2021). Design thinking: A creative approach to problem solving. *Management Teaching Review, 6*(2), 123–140.

Gloppen, J. (2009). Perspectives on design leadership and design thinking and how they relate to European service industries. *Design Management Journal, 4*(1), 33–47.

Hennessey, E., & Mueller, J. (2020). Teaching and learning design thinking (DT). *Canadian Journal of Education/Revue canadienne de l'éducation, 43*(2), 498–521.

Henriksen, H., Richardson, C., & Mehta, R. (2017). Design thinking: A creative approach to educational problems of practice. *Thinking Skills and Creativity, 26,* 140–153. https://doi.org/10.1016/j.tsc.2017.10.001.

Igbal, R. (2011). Impact of organisational change to achieve competitive edge *European Journal of Business and Management, 3*(4), 87–95.

Jamal, T., Kircher, J., & Donaldson, J. P. (2021). Re-visiting design thinking for learning and practice: Critical pedagogy, conative empathy. *Sustainability, 13*(2), 964.

Jaskyte, K., & Liedtka, J. (2022). Design thinking for innovation: Practices and intermediate outcomes. *Nonprofit Management and Leadership, 32*(4), 555–575.

Kahn, K. (2018). Understanding innovation. *Business Horizons, 61*(3), 453–46. https://doi.org/10.1016/j.bushor.2018.01.011

Kim, Y. S., & Park, J. A. (2021). Design thinking in the framework of visual thinking and characterization of service design ideation methods using visual reasoning model. *The Design Journal, 24*(6), 931–953.

Knight, E., Daymond, J., & Paroutis, S. (2020). Design-led strategy: How to bring design thinking into the art of strategic management. *California Management Review, 62*(2), 30–52.

Kolko, J. (2015). Design thinking comes of age. *Harvard Business Review, 93*(9), 66–71.

Lewis, J. M., McGann, M., & Blomkamp, E. (2020). When design meets power: Design thinking, public sector innovation and the politics of policymaking. *Policy & Politics, 48*(1), 111–130.

Liedtka, J. (2015). Perspective: Linking design thinking with innovation outcomes through cognitive bias reduction. *Journal of Product Innovation Management, 32*(6), 925–38.

Liedtka, J. (2018). Why design thinking works. *Harvard Business Review, 96*(5), 72–79.

Liu, A., & Lu, S. (2020). Functional design framework for innovative design thinking in product development. *CIRP Journal of Manufacturing Science and Technology, 30,* 105–117.

Luka, I. (2014). Design thinking in pedagogy. *The Journal of Education, Culture, and Society, 5*(2), 63–74.

Lynch, M., Kamovich, U., Longva, K. K., & Steinert, M. (2021). Combining technology and entrepreneurial education through design thinking: Students' reflections on the learning process. *Technological Forecasting and Social Change, 164*, 119689.

Magistretti, S., Ardito, L., & Messeni Petruzzelli, A. (2021). Framing the microfoundations of design thinking as a dynamic capability for innovation: Reconciling theory and practice. *Journal of Product Innovation Management, 38*(6), 645–667.

Malhotra, N., Zietsma, C., Morris, T., & Smets, M. (2021). Handling resistance to change when societal and workplace logics conflict. *Administrative Science Quarterly, 66*(2), 475–520.

McLaughlin, J. E., Wolcott, M. D., Hubbard, D., Umstead, K., & Rider, T. R. (2019). A qualitative review of the design thinking framework in health professions education. *BMC Medical Education, 19*(1), 8.

Mueller-Roterberg, C. (2018). *Handbook of design thinking.* Independently published..

Muratovski, G. (2021). *Research for designers: A guide to methods and practice.* SAGE Publications.

Nakata, C. (2020). Design thinking for innovation: Considering distinctions, fit, and use in firms. *Business Horizons, 63*(6), 763–772.

Nakata, C., & Hwang, J. (2020). Design thinking for innovation: Composition, consequence, and contingency. *Journal of Business Research, 118*, 117–128.

Noh, S. C., & Karim, A. (2021). Design thinking mindset to enhance education 4.0 competitiveness in Malaysia. *International Journal of Evaluation and Research in Education, 10*(2), 494–501.

Ortiz, R. (2016). IDEO new product development and human-centered business solutions.

Pande, M., & Bharathi, S. V. (2020). Theoretical foundations of design thinking–A constructivism learning approach to design thinking. *Thinking Skills and Creativity, 36*, 100637.

Panke, S. (2019). Design thinking in education: Perspectives, opportunities and challenges. *Open Education Studies, 1*(1), 281–306.

Randhawa, K., Nikolova, N., Ahuja, S., & Schweitzer, J. (2021). Design thinking implementation for innovation: An organisation's journey to ambidexterity. *Journal of Product Innovation Management, 38*(6), 668–700.

Rusmann, A., & Ejsing-Duun, S. (2022). When design thinking goes to school: A literature review of design competences for the K-12 level. *International Journal of Technology and Design Education, 32*(4), 2063–2091. https://doi.org/10.1007/s10798-021-09692-4

Thakur, A., Soklaridis, S., Crawford, A., Mulsant, B., & Sockalingam, S. (2021). Using rapid design thinking to overcome COVID-19 challenges in medical education. *Academic Medicine, 96*(1), 56–61.

Thawani, R. (2023). *Optimizing and Streamlining the Study Abroad Process using Design Thinking Methodologies and UX Strategies* (Doctoral dissertation, University of Toronto (Canada)).

Verganti, R., Dell'Era, C., & Swan, K. S. (2021). Design thinking: Critical analysis and future evolution. *Journal of Product Innovation Management, 38*(6), 603–622.

Waidelich, L., Richter, A., Kölmel, B., & Bulander, R. (2018). Design thinking process model review. *2018 IEEE International Conference on Engineering, Technology and Innovation (ICE/ITMC)*, Stuttgart, Germany, pp. 1–9. https://doi.org/10.1109/ICE.2018.8436281.

2 Pursuing an Innovation

Sixto Cabrera

2.1 INTRODUCTION

The pursuit of innovation refers to the systematic approach to identifying, developing, and implementing new ideas, products, or processes that enhance user experience and meet market demands (Brown, 2009; Tidd & Bessant, 2018). Innovation is at the heart of engineering, driving the development of solutions that push the boundaries of what is possible in product design. For any new product to be successful, it must provide clear benefits to the user experience, and these benefits should be measurable and supported by data. The process of pursuing innovation starts with identifying unmet needs or opportunities for improvement in existing designs (Christensen, 1997). Engineers must balance the desire for progress with the constraints posed by real-world factors such as cost, size, weight, and performance (Pahl & Beitz, 2007; Friedman, 2009).

In product design, especially within complex engineering fields, success is not simply a matter of inventing new features; it is about solving problems in ways that enhance the end user's experience while remaining practical and feasible (Brown, 2009). This requires a structured approach – one that evaluates the feasibility of improving current design limitations, considers the necessary trade-offs, and outlines a validation plan to prove the worth of the innovation (Montgomery, 2017). This chapter explores the step-by-step process that engineers should follow to identify, quantify, and implement innovative solutions, ensuring that any proposed changes are both beneficial and sustainable.

2.2 INNOVATION FRAMEWORK

One of the most intellectually rewarding exercises for an engineer is to conceive solutions to design puzzles. Most of the time the engineering challenges naturally arise from constraints imposed by demands of the final user application: size, cost, weight, performance, and so on. Considering the case of any battery-powered device, ultimately, energy must be efficiently utilized to prolong its usage. But any curious engineer may ask: *What are the factors that influence how long the battery will last? What trade-offs can be made to favor battery life? Why is a battery needed in the first place?* These are the kinds of questions that foster design thinking. It is about questioning the status quo, maximizing limited resources and constraints, understanding that benefits come at a cost, and that the most important metric is always the impact to the final user (Christensen, 1997; Ulrich & Eppinger, 2012). The pursuit of an innovation starts with identifying a need that has not been addressed with the current design solution – this is referred

TABLE 2.1
Innovation Framework

	Innovation Framework	Example: Battery-Powered Outdoor Camera
1	What is the benefit over the baseline?	Enhanced wireless range to 180 ft; baseline supports up to 100 ft
2	How is the benefit measured or quantified?	Average range increase in feet
3	What are the limiting factors to the baseline design?	Radio frequency (RF) transmitter power limited to 12 dBm No more room in thermal budget
4	What are the trade-offs in exchange for the benefit?	Battery life impact Increase product dimensions

to as the *benefit*. Then, such benefit must be measured and compared against the incumbent or baseline. Next, it is necessary to identify what are the limiting factors to the current baseline design. And lastly, the benefit will always come at a cost, thus it is critical to understand the trade-offs and assess whether the benefit outweighs the cost (Montgomery, 2017).

Extending the battery-powered device example, let us assume that the product is an outdoor camera, and one feature users have shown desire for is enhanced wireless range to allow the camera to be placed farther away from their house.

Even though a profitability assessment is not listed in the innovation framework (Table 2.1), the innovation must provide a reasonable return on investment to the company, otherwise it may not be supported by leadership (Cooper, 1990; Christensen, 1997). For simplicity, let us assume, in this chapter, that the innovation has been proven to be profitable and thus all that is left to establish are the technical benefits.

2.3   KEY DATASETS

Today, business and engineering decisions are heavily data-driven; this implies that the benefit being pursued needs to be supported by evidence which are referred to as *key datasets* (Ulrich & Eppinger, 2012). It is worth noting that these key datasets not only need to corroborate the benefit achieved by the design choices but also need to show that no new side effect(s) are created such that the innovation is not worth the effort. Even though a few trade-offs may be easy to predict, as those listed in the innovation framework (see Table 2.1), it is likely that unintended consequences may arise because of unaccounted variables or process variations, making the need for actual data very important.

Design thinking requires taking a holistic view of the project. Naturally, this poses further challenges as it requires domain knowledge to make assessments about how design changes may impact other product areas beyond one's own expertise (Hargadon, 2003). To address the gap, it is necessary to leverage cross-functional teams. One efficient way is to request cross-functional teams' specific key datasets that, from their perspective, are needed to confirm that there are no major side effects (Shmueli et al., 2010; Ulrich & Eppinger, 2012). Let us assume that for the battery-powered outdoor camera, the relevant teams are radio frequency (RF), battery, analog, electromagnetic compliance (EMC), Software (SW), mechanical, and reliability. While it is possible that some of the key datasets show no adverse impact at all, it is also possible that others do show some side effects. The question then would be whether the side effects are an acceptable trade-off for the innovation.

Going back to our battery-powered outdoor camera, it makes sense to think that if more electrical power is required to improve wireless range, then battery life will be impacted, unless the battery is replaced with a more capable one (with additional trade-offs). There are other potential impacts, with associated risk levels, that cross-functional teams may highlight as indicated in the Table 2.2.

One important element to keep in mind is that *not all* the datasets are must-haves. Generally, if the team can decide whether to proceed with the innovation efforts without a specific dataset, such dataset is not a must-have. Let us assume that the battery-powered outdoor camera's advertised operating temperature range is −20°C to 65°C. There may still be a request from the reliability team to qualify the design at 90°C to assess aging, durability, and performance under such harsh conditions. While this is great data to estimate margin to failure, it should not necessarily prevent proceeding with the innovation if failures were observed beyond the 65°C upper limit. This is still a nice-to-have dataset as it may indicate other regressions that may or may not be worth looking into when compared with the baseline design.

TABLE 2.2
Cross-functional Key Dataset Dashboard

			Risk Level	Low	Medium	High
Team	**Risk**	**Key Dataset(s)**	**Rationale**			
RF		• Wireless range • Tx Quality	Need to confirm range improvement and validate no Tx signal quality regression			
Battery		• Avg. Power increase	Battery life will reduce 1 day per 1mW increase			
Analog		• Residual noise	More Tx power could lead to noise coupling to analog section making it harder to distinguish real vs noise signal			
EMC		• 6x samples needed for regulatory tests	Risk to fail regulatory standards due to harmonics as a result of higher Tx power			
SW		• 2x samples to check bit-error rate	Higher power could produce signal/power integrity issues and thus link bit error rate increase			
Mechanical		• 2x samples for thermal/mechanical stress tests	Design change will increase heat dissipation and physical size			
Reliability		• 5x samples for 65C & 85C High Temp Operational Test	Need to validate design robustness in harsh conditions			

Abbreviations: EMC, electromagnetic compliance; RF, radio frequency; SW, Software.

2.4 DESIGN OF EXPERIMENTS

Once the key datasets are identified, a design of experiment (DoE) is developed. This is the vehicle used to collect and analyze the impact of the innovation (Montgomery, 2017). Simply put, a DoE is a small production run that implements the design changes necessary to enable the pursued benefit. The DoE has three main purposes: (1) verify that the benefit can be realized; (2) quantify the impact of the side effects, especially the unexpected ones; and (3) understand the inherent production *process variation* and how it impacts the innovation.

One of the first considerations is determining the number of samples needed to make good inferences. The bottom line is that there are no strict rules and companies have their own criteria rooted in statistical insights, product specifications, and costs. Let us assume the company produces 5,000 units per year and has empirically determined that a 2% sample size on the yearly production is needed to qualify any potential innovation. For the battery-powered outdoor camera example, this implies that the DoE sample size will be 100 units. Something to keep in mind is that while the sample size may be sufficient to reveal the average performance, from a design thinking perspective there are additional layers that must be considered. Chances are that *corner* cases may not be captured. In engineering, a corner case involves a problem or situation that occurs only outside normal operating parameters – specifically one that manifests itself when multiple environmental variables or conditions are simultaneously at extreme levels, even though each parameter is within the specified range for that parameter. Imagine that to enhance the outdoor camera's wireless capability, a new RF amplifier integrated circuit (IC) chip is required. From silicon level fabrication process, the IC chips may very well fall within a few performance categories depending on the underlying characteristics of the transistor technology.

Without going into the complexities of silicon process, the IC chip is an arrangement of N-type and P-type transistors. Now each transistor type, either N or P, can be "fast" or "slow," or something in the middle, which can be considered "typical." Based on this, several two-dimensional (2D) permutations arise depending on the characteristics of the N- or P-type transistors: *Typical (TT), fast-fast (FF), fast-slow (FS), slow-fast (SF), and slow-slow (SS).* This is illustrated in Figure 2.1

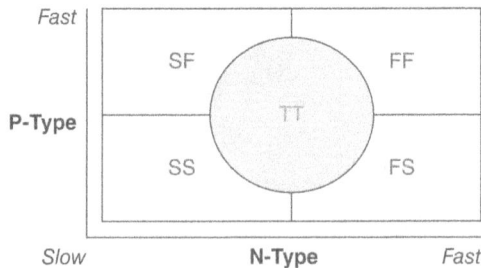

FIGURE 2.1   Silicon process table.

TABLE 2.3
DoE Quantity Breakdown

	DoE Breakdown	Quantity
1	Typical (TT)	1,000+ (from existing production run)
2	Fast N-type and fast P-type (FF)	20×
3	Slow N-type and slow P-type (SS)	20×
4	Fast N-type and slow P-type (FS)	20×
5	Slow N-type and fast P-type (SF)	20×

in which each quadrant represents the different combinations, with typical part centered in the chart. For the battery-powered outdoor camera, this implies that the RF amplifier needs to boost the wireless range and depends on whether the part is fast, slow, or typical. If the company is targeting to build 5,000 outdoor camera units per year and each one needs 1× RF amplifier IC, it is likely that all these kinds of IC chips (TT, FF, FS, SF, and SS) will be in the production mix. *Part of design thinking is now to understand the implications of each kind in the IC chip distribution list: How will the different corners impact the range? What about power consumption?*

Armed with this information, the DoE sample size quantity can be further split up to purposefully include corner cases. For example, the IC chip manufacturer can be asked to provide corner parts for validation purposes as shown Table 2.3.

A valid question at this point is, if specific corner parts can be identified, isn't it possible to simply screen corner parts out in the first place? In mass production, it is just too costly to control the fabrication process in such a way. First, screening out parts for being a corner means that production yield will take a hit, and second, over time process deviations may still introduce corner cases that somehow escape any screening. The implication of this is that during the design thinking phase, corner cases should be taken into account to ensure the innovation can still be achieved under all plausible circumstances. The same idea of IC chips corner cases can be extrapolated to other aspects of the design changes associated with the innovation.

2.5   TIMELINES

In real business dynamics, products are launched on a regular basis depending on the market a company operates in. For example, the big smartphone companies release new products every year as they battle to grab more market share. Imagine that a new battery-powered outdoor camera is launched every two years. This implies that the DoE results should time out before any critical decisions need to be made, otherwise the company will be taking on additional risks if aspects of the design are not fully validated and the company ends up rushing to market.

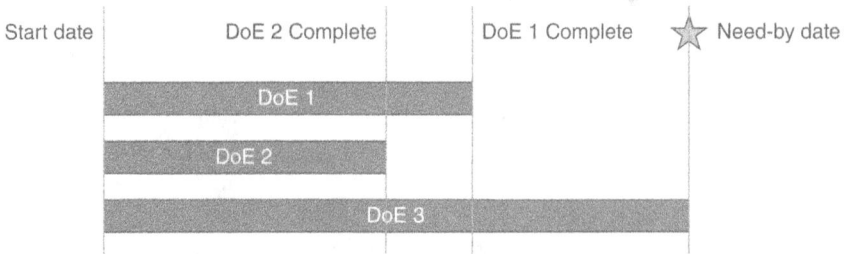

FIGURE 2.2   Right-to-left schedule timeline.

Pursuing an innovation is not only about designing a feature that adds value
to a product, but also about thinking how the innovation will intercept with the
company's timeline to launch. One way to plan accordingly for any DoEs is to
calculate backward from the need-by date and that will dictate how late the DoE
can be kicked off and when the key datasets will be available in time (Cooper,
1990; Ulrich & Eppinger, 2012). This is the concept of a *right-to-left* schedule
(Figure 2.2). The figure simply shows the need-by date on the far right and when
the DoE results will become available. Notice that in this example, "DoE 3" takes
longest amount of time and lines up exactly with the need-by date, which could
pose delay concerns as there is no room for schedule slips.

In practice, the key datasets may still be incomplete by the need-by date
due to a wide range of roadblocks such as process delays, equipment issues,
wrong test conditions, etc. In such a situation, the team needs to assess if the
limited dataset is enough to make a risk call. That said, design thinking should
also be proactive instead of reactive. What this means is that if one of the key
datasets is trending as if it will miss the target date, it may be time to consider
what to prioritize within the key dataset to speed up the results. For example, if
the RF team initially requested wireless transmission quality (Tx) at 12 different power levels, an option to reduce test time could be to check Tx quality at
fewer power levels (e.g., 6). While this implies a loss in resolution, it may still
be good enough data points to conclude if there are any side effects to the new
RF IC chip.

2.6   ASSESSING DOE RESULTS

For the outdoor camera example, DoE results from the RF, battery, and analog
teams are presented in the boxplots shown in Figures 2.3–2.6).

Notice how the datasets in Figures 2.3 to 2.6 are broken up per RF IC chip
corner type to understand the impact coming from each one. Another thing to
keep in mind is that the baseline dataset does not necessarily have to be the
same size as the DoE. Ideally, the company should have a large amount of
baseline data as units are currently being produced. The importance of a large
baseline dataset is that it allows for high confidence levels where the current
(and historical) production process is, meaning that if there were deviations

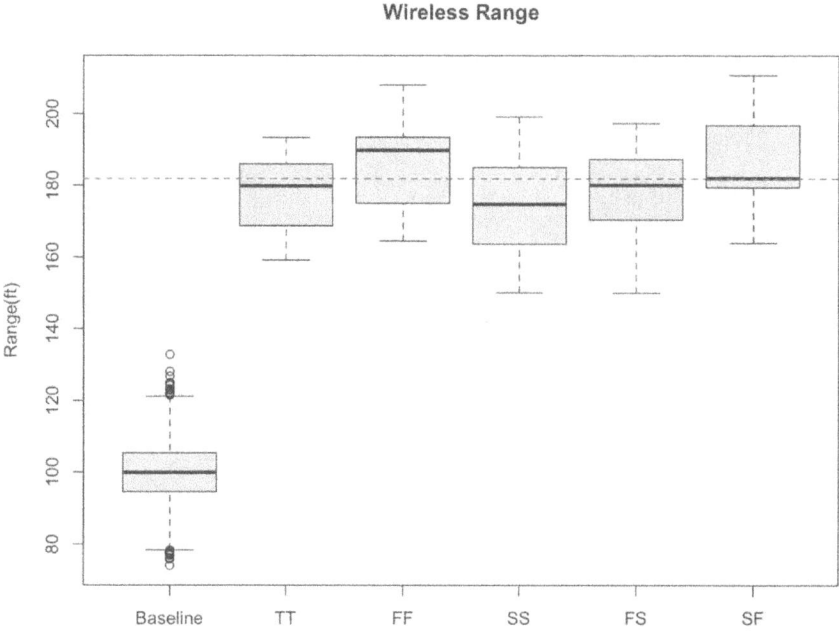

FIGURE 2.3 Wireless range boxplot.

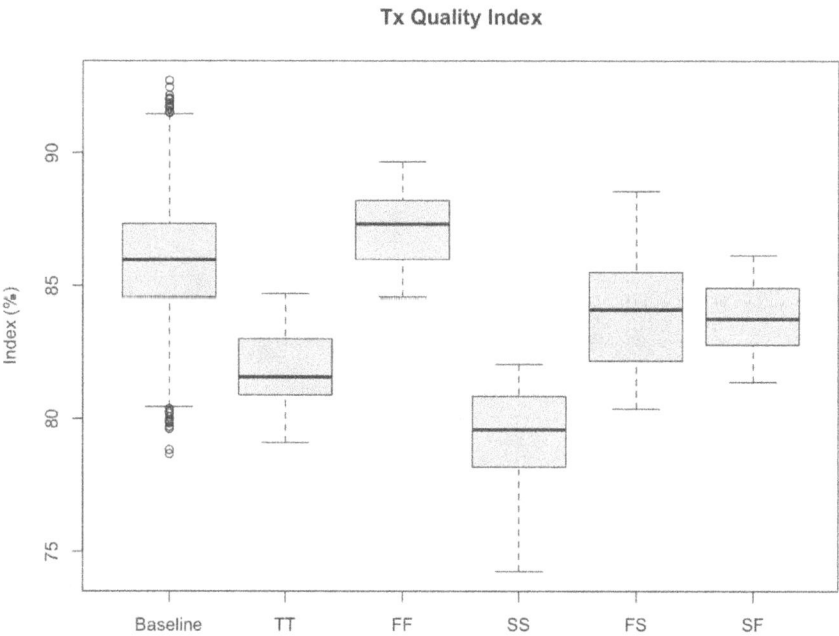

FIGURE 2.4 Tx quality boxplot.

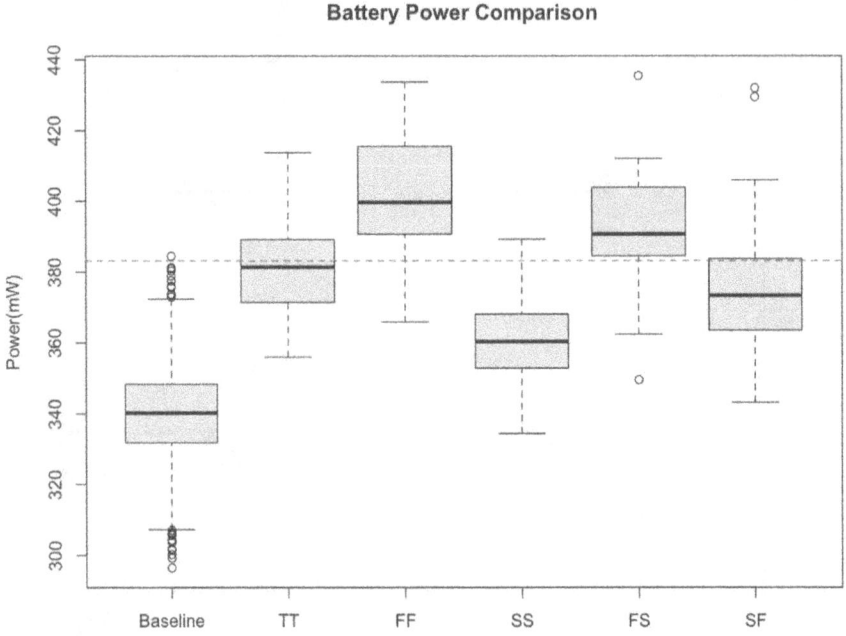

FIGURE 2.5   Battery power comparison boxplot.

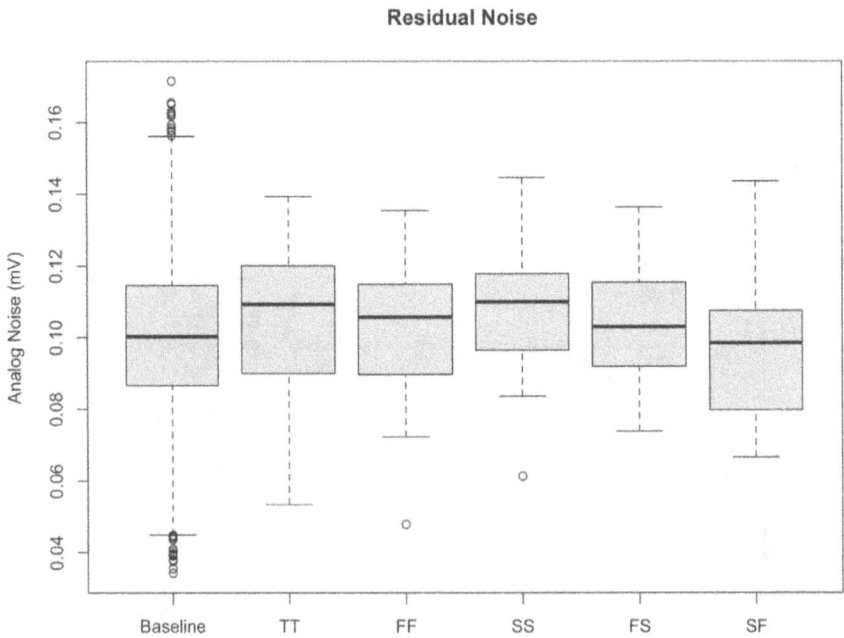

FIGURE 2.6   Analog noise boxplot.

in the DoE results, greater than the baseline process spread, such deviation would most likely come from a true characteristic of the DoE. The DoE results clearly show that the 180 ft wireless range is achievable with the new RF IC chip (Figure 2.3). It is worth noting that while the average range is indeed 180 ft, and the distribution is such that there are outlier units hovering around 150 ft, some sit 210 ft. From an engineering design thinking perspective, this is the opportunity to reflect on the variables that could be driving this spread: *Is it due to part-to-part calibration process? Is it an inherent characteristic of the new RF IC chip? Is there a way to optimize our process to skew the distribution toward the larger range extreme?*

Figure 2.4 shows the Tx index. Let us assume that the minimum acceptable index is 70%. Then the SS corner type may have outliers close to failing this benchmark. Whether the RF team will sign off on the innovation, in light of the SS marginal performance, needs to be determined but it opens the conversation about next steps to address the concerns. The best course of action is trying to understand the fundamental root cause that makes the SS corner marginal, but it may also be a good idea to challenge the status quo and understand why 70% is deemed the minimum acceptable Tx quality index. *Is the team over-designing? What is the user impact if Tx quality index were something lower than 70% (e.g., 65%)?*

Another critical result to pay attention to is the battery power increase as a trade-off for the enhanced range. The data shows an average 40-mW power uptick across the entire DoE with respect to the baseline (Figure 2.5). In particular, for the FF corner type, the power increase can amount to up to approximately 100 mW. From empirical knowledge the company knows that every 1 mW increase will cost 1 day of battery life. Thus, if the expected battery life is 3 years, based on the FF corner, 100-mW increase will translate into 100 days of battery life loss. Whether this has a significant impact or not depends on how the user perceives it. From a design thinking perspective, the goal is to still gain back the loss and for that it is imperative to rethink the variables that impact battery life. *What if another feature of the outdoor camera were disabled or dialed down? Could that recover some of the loss in battery life?*

Finally, the DoE results also show no impact to the analog noise performance (Figure 2.6). This is welcome news, but it is important to obtain final assessment from the corresponding teams both to confirm nothing has been overlooked and that the innovation poses no concerns to them

2.7   MAKING THE CASE FOR AN INNOVATION

An essential skill for an innovation is the ability to abstract complex information and present it in simple terms that are easy to grasp, even for non-technical audiences (Rappaport, 2002). This skill not only helps get the message across to other people but reinforces one's own understanding of the problem statement, the findings, and the conclusion that needs to be conveyed.

One way to efficiently summarize the salient points is by creating a single presentation slide or *one-pager* with the following three sections (see Figure 2.7):

- *Key takeaway:* In one sentence, what is the one idea that the audience needs to take away from the innovative work.
- *Findings:* Summarize the results (e.g., bullet points) from the DoE and state how the observations support the innovation or not.
- *Next steps:* What information is still pending or missing? What are the next steps to close any gaps?

Pursuing an Innovation – Outdoor Camera Range Improvement

Key Takeaway
❑ Enhanced wireless range to 180 feet is achievable without significant user impact ❶

Findings
❑ Worst case battery life reduction by 100 days but still tolerable
❑ Tx Quality and Analog noise figure of merits remain within specs
❑ EMC results show passing FCC & CE regulatory results

Next Steps
❑ Scope power reduction other camera subsystems, *ETC: 1 week*
❑ Validate SW processing time increase will not adversely impact functionality, *ETC: 2 weeks*

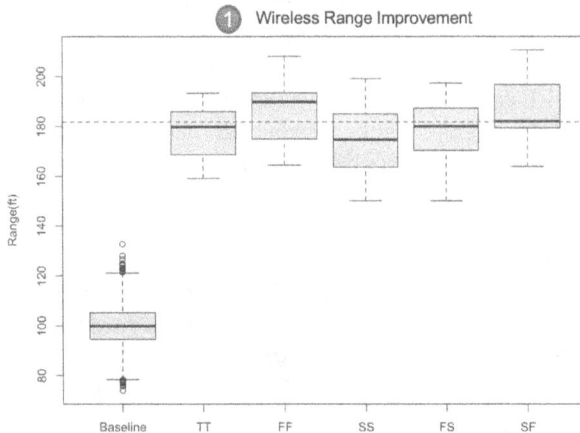

FIGURE 2.7   Example one-pager.

Reflecting on the battery-powered outdoor camera example, the DoE results strongly show that enhanced wireless range up to 180 ft is achievable without any side effects. While there are concerns about decreased battery life as a trade-off, the user impact is still considered low. With that in mind, the final message or *key takeaway* needs to be communicated concisely. In the *findings* section, the idea is to connect the DoE results to the key takeaway. Based on the DoE results, it can be stated whether the observations suggest or support some conclusion. The EMC team's tests (Table 2.2) clearly showed that the product will still comply with regulatory standards. There are cases where the observations suggest a conclusion, but the connection is not as strong as other DoE results. These open questions must be followed up on and reflected in the next steps section. For example, the SW team indicated (Table 2.2) that while there were no integration problems with the new RF IC chip in the outdoor camera, one observation was that computation processing time increased with respect to baseline. This may require SW workaround in the future if a new firmware update does not work with the increased timing and introduce race conditions.

An example of what a one-pager may look like based on the recommendations provided in this chapter is illustrated in Figure 2.7. While a one-pager should be able to capture the key message, it is important to have backup documentation in case questions are raised and more details are needed. Examples of additional information are decision trees, hypothesis tables, cross-functional team risk assessments, and anything that adds value and strengthens the case for the innovation.

2.8   CONCLUSION

The pursuit of innovation in engineering is both challenging and rewarding, requiring a careful balance between creativity and practicality. It is a journey that begins with identifying a need or opportunity for improvement, followed by a rigorous process of validation, data analysis, and then communicating the validation findings with the key stakeholders (Figure 2.8). While the process involves substantial collaboration across multiple teams and disciplines, it is this very complexity that drives meaningful design thinking. At its core, innovation is about enhancing the end user's experience while navigating the constraints of cost, performance, and feasibility. The key to success lies in not only achieving technical benefits but also ensuring that those benefits justify the associated costs (Rappaport, 2002). Furthermore, the process demands a forward-thinking mindset – one that anticipates potential challenges and continuously adapts to evolving project goals (Ulrich & Eppinger, 2012). As engineers, the ultimate goal is to deliver a solution that provides tangible value, supported by data and validated through thoughtful experimentation (Montgomery, 2017). By embracing this structured approach to innovation, engineers can push the boundaries of design and contribute to the creation of products that truly make an impact.

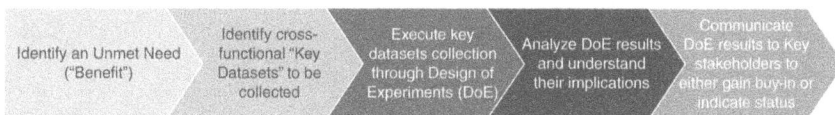

FIGURE 2.8   Steps for pursuing innovation by identifying needs, collecting data through experiments, analyzing results, and sharing outcomes with stakeholders

REFERENCES

Brown, T. (2009). *Change by design: How design thinking creates new alternatives for business and society.* Harper Business.

Christensen, C. M. (1997). *The innovator's dilemma: When new technologies cause great firms to fail.* Harvard Business Review Press.

Cooper, R. G. (1990). Stage-gate systems: A new tool for managing new products. *Business Horizons, 33*(3), 44–54. https://doi.org/10.1016/0007-6813(90)90040-I

Friedman, G. J. (2009). *The design of engineering systems: Models and methods.* Wiley.

Hargadon, A. (2003). *How breakthroughs happen: The surprising truth about how companies innovate.* Harvard Business School Press.

Montgomery, D. C. (2017). *Design and analysis of experiments* (9th ed.). Wiley.

Pahl, G., & Beitz, W. (2007). *Engineering design: A systematic approach* (3rd ed.). Springer. https://doi.org/10.1007/978-1-84628-319-2

Rappaport, T. S. (2002). *Wireless communications: Principles and practice* (2nd ed.). Prentice Hall.

Shmueli, G., Patel, N. R., & Bruce, P. C. (2010). *Data mining for business intelligence: Concepts, techniques, and applications in Microsoft Office Excel with XLMiner.* Wiley.

Tidd, J., & Bessant, J. (2018). *Managing innovation: Integrating technological, market and organizational change* (6th ed.). Wiley.

Ulrich, K. T., & Eppinger, S. D. (2012). *Product design and development* (5th ed.). McGraw-Hill.

3 Designing Ideas

A Dual Perspective

Jana Huck and Alexandre Sukhov

3.1 INTRODUCTION

Developing new ideas is central to design, since designers are often put in a situation where they face a practical problem and are asked to generate insights for potential solutions. During idea development, the focus is often on modifying and improving ideas through knowledge integration and creative work (Obstfeld, 2005; Tsoukas, 2009). However, for ideas to have a better chance of being selected for implementation, idea development may include engaging in and managing social interactions within the organization to gain internal support (Perry-Smith & Mannucci, 2017; Cohendet et al., 2024). Although both the sides of idea development are important, current research lacks a holistic understanding on what it means to effectively manage idea development. One of the reasons for this is that different literatures use fundamentally different assumptions on how they approach the core concept of an 'idea' and idea development (Hua et al., 2022). Each approach has significant implications on what constitutes a 'good idea' from each of the perspectives, and what types of idea development activities are important.

Ideas can be broadly defined as "provisional and communicable representations" by Hua et al. (2022, p. 18), which means that they are bound to change; are shared through verbal, written, or visual means of communication; and stand for something that has not yet materialized (Hua et al., 2022). This definition illuminates the main characteristics of what makes an idea and hints at their complex, somewhat unstable, and rather subjective nature.

Furthermore, previous research describes ideas mainly in two ways. One way is to view ideas as *objects*, where designers focus their problem-solving skills and use methods for developing desirable (Liedtka et al., 2015), viable (Seidel & Fixson, 2013), and feasible (Brown, 2008; Micheli et al. 2019) distinct ideas as the main outcome of ideation and idea development process (; Dorst & Cross, 2001; Dean et al., 2006; Girotra et al., 2010). Another way is to view ideas as *triggers-for-change*, where designers focus more on the reactions that the ideas set in motion, such as, the exploration of new meanings and creating a drive for ideas to move forward. This emphasizes the importance of facilitating supporting environment for ideation (Perry-Smith & Mannucci, 2017; Sosa, 2019; Sukhov et al., 2019; Sukhov et al., 2021; Hua et al., 2022). The perspective adopted by designers and innovation managers to understand ideas will influence how idea development and the design process will be organized, what

DOI: 10.1201/9781003487524-3

activities will be considered important, and what outcomes of this process will be most desired (Sosa, 2019).

Although these perspectives are helpful for identifying different types of support activities for idea development, each perspective is also inherently limited in understanding the complexity of what is happening during idea development, making it difficult to give clear suggestions on how to further improve the idea development process. The purpose of this chapter is to provide additional clarity to what these perspectives entail and how they can help designers and innovation managers to navigate the complex and dynamic idea development process. In the following sections, we delve deeper into each perspective and outline the key aspects of each perspective enabling managers to understand what can improve idea development process in their organizations.

3.2 IDEAS AS OBJECTS

Traditionally, in the innovation management literature, ideas have been understood as objects or entities that are the outcomes of focused problem-solving activities (Osborn, 1957; Sukhov et al., 2019), embryos for a product (Kahn et al., 2013), or creative products (Amabile, 1982, 1998). From this perspective, the goal of research on the early stage innovation process has often been to find the best ideas (Hammedi et al., 2011), identify the best source of ideas (Antons & Piller, 2015), and improve the efficiency of this process (Cooper & Kleinschmidt, 1987; Cooper, 1988; Eling et al., 2015). One way to portray this process is how problem-solution objects pass through a funnel (Figure 3.1). The process starts with many ideas and in the end the best idea is selected for implementation (Koen et al. 2001).

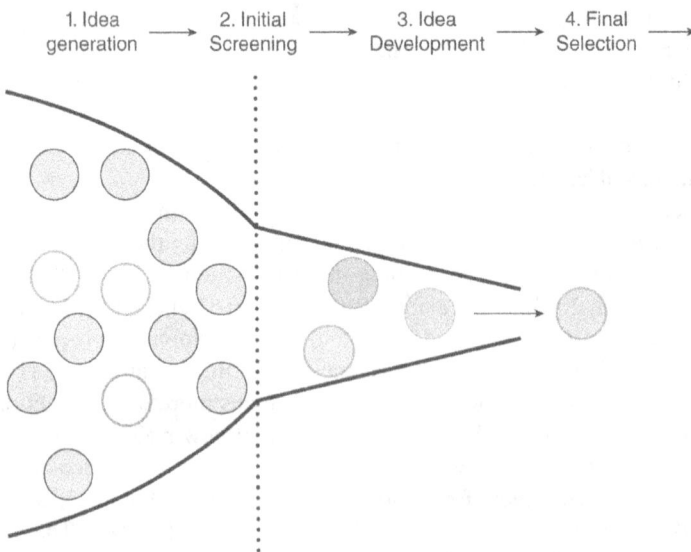

| 1. Idea generation | 2. Initial Screening | 3. Idea Development | 4. Final Selection |

FIGURE 3.1   Ideas as objects moving through the idea funnel.

Given that ideas are *provisional*, i.e., they are bound to change, the focus of idea development, from the idea-as-objects perspective, is to turn low-quality ideas into high-quality ideas by focusing on various aspects where the ideas are lacking (Florén & Frishammar, 2012). To do so, ideas are usually elaborated, clarified, and expanded (Beretta, 2019), with the aim of reducing uncertainty and ambiguity (Cooper, 1990), by collecting new information and integrating new knowledge (Florén & Frishammar, 2012). This can take different forms. A common one is to involve different stakeholders during idea development to review and complement/co-create the idea by editing it with the help of their expertise in different areas. This process can be described as the co-evolution of the initial problem and solution description until the final idea concept is reached (Maher et al., 1996; Dorst & Cross, 2001). By introducing different people and viewpoints on the idea, new knowledge can enter the idea development process. This helps to overcome the fixation with the initial idea concept (Dane, 2010; Boudier et al., 2023) and helps to realize new ways of looking at the initial problem (Hatchuel & Weil, 2009).

Ideas that are understood as objects can be expressed in verbal, written, or other visual forms (Sosa, 2019), and for something to be considered a 'good idea' it is important that it is clearly *communicated* and easily understood (Hua et al., 2022), since lack of clarity has been found to reduce perceived idea quality (Sukhov, 2018). It is also common to focus on reducing cognitive biases during idea evaluation and selection (Licuanan et al., 2007) to ensure that in fact the best ideas are moved into implementation and not the decision-makers' personal agendas, or other forms of personal constraints (Dane, 2010; Girotra et al., 2010; Sukhov, 2018). One way to reduce personal biases during idea evaluation is to use a panel of independent raters that are familiar with the idea's domain and identify ideas that these raters agree on by using consensual assessment technique (CAT) (Hennessey, 1994). Another way is to improve idea clarity and overcome the potentially biased personal interpretation by improving an idea's completeness (Sukhov, 2018), elaboration (Beretta, 2019), or specificity (Dean et al., 2006). This can be achieved by explaining the what, why, and how a certain problem can be resolved (Karlsson & Törlind 2016; Sukhov et al., 2019) or similarly using the five-whys technique to get to the root cause of a particular problem. There are of course other approaches that can help in idea development, e.g., the TRIZ approach (theory of inventive problem-solving). TRIZ offers various tools to identify the current problem (current state), identify available resources, determine the goals that should be fulfilled by the solution (the intended state) and map the steps to transition from current to intended state (Moehrle 2005; Marcandella et al. 2009).

Furthermore, for ideas-as-objects to have high quality, they need to act as well-defined and worked-out *representations* that can be clearly distinguished from other ideas. Idea quality is often evaluated by external judges against criteria such as relevance, use value, workability, feasibility, specificity, and novelty (Dean et al., 2006; Girotra et al., 2010; Magnusson et al., 2016). This helps to prioritize among tens, hundreds, or even thousands of ideas by distinguishing between different types of ideas that hold potential as future innovations. Designers can also use these criteria to direct their idea development activities to enhance ideas in a specific way, e.g., make ideas more feasible given the resources available, or improve solutions to problems with respect to the value it offers to target users, or help identifying the resources necessary for implementation (Sukhov et al., 2019). In practice, designers can use idea templates (Naggar 2015) that are aimed at expanding the content of the idea and clarifying its context through development of problem and solution components (Sukhov et al., 2019). Idea templates are similar to a business model canvas (Osterwalder & Pigneur, 2010) and can help designers to clearly explain ideas with regards to the relevant evaluation criteria. Organizations can further use idea quality criteria to prioritize the development of ideas that are easy to implement, ideas that have high market value or ideas that are new to the market. Using evaluation criteria can help build an ideation port- folio for different types of innovations and better allocate the resources needed for their development (Kock et al. 2015). Easy to implement ideas, which involve incremental changes, often require minimal effort or resources, while potential high market value ideas, such as developing and training AI for medical diagnos- tics and drug discovery, might require more long-term investments and efforts.

By treating ideas as distinct objects, they are given an independent existence, making it possible for them to be owned and transferred both psychologically and legally (Hua et al., 2022). Psychological ownership involves a person per- ceiving an idea or entity as 'ours' or 'mine' (Baer & Brown, 2012). This sense of ownership influences the idea development process by affecting how feedback is received and how new knowledge is integrated. High degrees of ownership might lead to the addition of new elements to the idea, while rejecting the removal of elements to which the owner is attached (Baer & Brown, 2012). This reluctance to 'kill their darlings' can inhibit the development of clear and concrete ideas.

Another important aspect is that ideas, being objects, can be transferred to others (Sukhov et al., 2019), making individual contributions to modifying ideas more evident. A common practice from this perspective is to incentivize idea creators by offering monetary rewards, recognition, and making it meaningful for participants to generate and submit ideas on ideation platforms and during idea challenges (Baruch et al., 2016). It is important to highlight that treating ideas as objects means they can be considered potential patents or exhibit property rights, leading to legal repercussions if ideas are 'stolen').

Overall, treating ideas as objects helps to trace ideas in the front end of inno- vation process, and identify the success factors for the best ideas, following the survival-of-the-fittest approach (Campbell, 1960; Simonton, 1999). From this per- spective, idea development activities can be understood as introduction of new information and knowledge that help to (i) enhance and adapt ideas to the orga- nizational goals and abilities to implement, (ii) concretize ideas to improve idea clarity and communicability, and (iii) transform ideas into new combinations of problem-solution pairs (Seidel, 2007; Boudier et al., 2023), where we can trace individual contributions that helped to modify and improve ideas. Taken together, idea development, from the idea-as-objects perspective, can be considered a

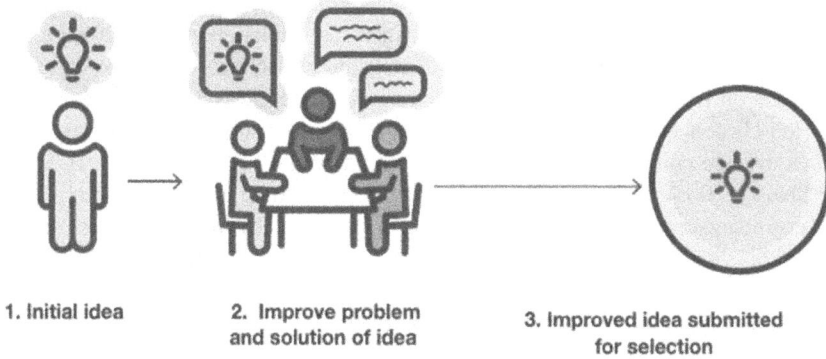

1. Initial idea   2. Improve problem and solution of idea   3. Improved idea submitted for selection

FIGURE 3.2   Idea development from an idea-as-objects perspective.

creative problem-solving process with an objective of improving ideas in terms of enhancing idea quality (Figure 3.2).

3.3   IDEAS AS TRIGGERS-FOR-CHANGE

While the idea-as-objects perspective focuses on enhancing idea quality through structured development and problem-solving, we are now introducing the triggers-for-change perspective, which shifts the focus towards the dynamic role of ideas in eliciting different, interpretations and actions towards change. This new viewpoint emphasizes the importance of engaging with ideas to stimulate discussions, create new associations, and drive creative behaviour.

Following the ideas as triggers-for-change perspective, ideas can be viewed as enablers for actions. As such, ideas can be described as triggers that create new associations (Sukhov et al. 2019; Boudier et al, 2023), stimuli for group discussion (Gillier & Bayus, 2022), acting as important drivers for change (Perry-Smith & Mannucci, 2017). From this perspective, the goal of the early-stage innovation process is not limited to finding the best ideas, but rather to realize the value that is created by engaging with ideas in various ways. For instance, it can be valuable for organizations to facilitate a continuous flow of ideas (Sosa, 2019), encourage flexibility and exploration of new paths (Boudier et al., 2023), and create support for driving the ideas forward (Perry-Smith & Shalley, 2003; Kijkuit & Van Den Ende, 2007), aiding in their implementation as new products, services and/or processes (Perry-Smith & Mannucci, 2017). Therefore, the focus of this perspective is on understanding the process and effects created by engagement with the ideas. The trade-off of this perspective comes with the limited resources that an organization has. This means that only a selected number of potentially interesting ideas can be viewed as triggers and be fully explored.

Viewing ideas as triggers-for-change, means that an organization emphasizes the *provisional* nature of ideas and can focus more on sustaining and directing the types of interactions that arise during idea development process (Perry-Smith & Shalley, 2003; Kijkuit & Van Den Ende, 2007). For instance, Gillier and Bayus (2022) found that the type of stimulus ideas in group discussions can influence how the stream of the ideation process unfolds. This view emphasizes that ideas act as fragments that are loosely connected to other fragments, similar to the dynamic social network structures that are commonly used to describe creativity and interactions (Perry-Smith & Shalley, 2003). Clusters of these fragments may build up through the interplay of divergent and convergent activities resulting in a defined design concept (Sosa, 2019). From this view, one of the most important factors to realize, is to help and facilitate the environment where these fragments of different knowledge can flourish.

From an idea as triggers-for-change perspective the key for an idea to exist, is that they are shared and *communicated* between different people to form new interpretations and attachments (Hua et al. 2022). A 'good idea' from this perspective is an idea that helps to realize new meanings (Verganti & Öberg, 2013), expand ideation (Boudier et al., 2023), improve confidence in decision-making (Sukhov et al., 2021), and helps to gain legitimacy (Perry-Smith & Mannucci, 2017). Ideally, ideas that are perceived as triggers, should help in creating and sustaining discussions, draw attention, be popular, and help in increasing people's motivation to work on the ideas further. The best idea is therefore not the one that offers the most optimal problem-solution, nor a clear and unambiguous description for a certain course of action, but the one that stimulates a bigger engagement and/or impact. This emphasizes that ideas are not existing in a vacuum, but they are constructed by people who can envision their potential, recognize the 'hidden' value, and able to take actions on helping the ideas to align with the organization to be realized. Thus, one of the main idea development activities is to improve sharing ideas with other people, activating social networks, and capture feedback, engagement and actions that ideas help to create (Ter Wal et al., 2022).

Even from ideas as triggers-for-change perspective, ideas *represent* something that is not yet materialized and are inseparable from the process that unfolds to improve them, rather than the outcome of the idea development process. Ideas can encompass fragments of information on how a certain problem can be solved, however, the content of the idea description is not as important compared to what the ideas can set in motion. Ideas can represent different things to different people working with them (Sukhov et al., 2021; Boudier et al., 2023). However, what is important is whether an idea is seen as 'promising' or 'interesting' in order to motivate further development and implementation (Sosa, 2019). Similarly, Hua et al. (2022) argue that it is essential for ideas to attract interest and support as it sparks the conversations that keep the design process going.

How ideas are developed can also act as a legitimization process, improving the idea's chances of being understood and accepted (Sukhov et al., 2021). As such, the boundaries of ideas are fluid and inseparable from the consequences they create (Hua et al., 2022). This means that depending on who works with or is exposed to an idea will hold different interpretations. The meaningfulness of the idea can only be realized upon interaction and engagement. By highlighting the sensemaking activities involved in idea development, more knowledge and more engagement with the ideas can be obtained (Sukhov et al., 2021). From the ideas as triggers-for-change perspective, ideas are schemas for action that cannot be transferred to others but can be understood in the interpretations and actions they create (Tsoukas, 2009). Idea ownership is a tricky concept to associate with this perspective, because in the legal sense it cannot be determined, whilst psychologically, people can be considered owners of their own actions that influence the ideas path in an implicit and shared sense (Nonaka, 1994). For instance, a team develops a marketing strategy centred on using social media influencers, with each member interpreting the idea through their own lens—sales, design, or R&D—and taking actions that shape its execution. The idea itself is not owned by any individual, but each person feels ownership over the actions and decisions they contribute to its realization. This collective process illustrates how ideas are schemas for action, understood through the interpretations and actions they inspire, rather than something that can be legally or singularly owned.

Relating to this is the notion of what makes a significant contribution to idea development and 'who' is involved. A significant contribution to idea development can be divided into two stages: first, identifying who provides feedback and initial support; and second, determining who can help the idea withstand external criticism (Mannucci & Perry-Smith, 2022; Ter Wal et al., 2022). From the ideas-as-triggers perspective, the focus is on the social process initiated, rather than the content of the idea itself (Figure 3.3) mainly as turning to others for

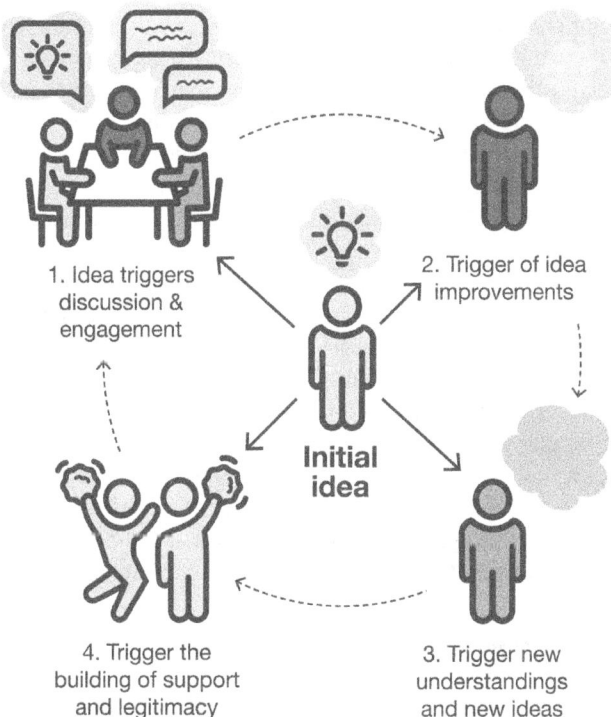

1. Idea triggers discussion & engagement

2. Trigger of idea improvements

Initial idea

4. Trigger the building of support and legitimacy

3. Trigger new understandings and new ideas

FIGURE 3.3   Ideas as triggers-for-change cause different reactions.

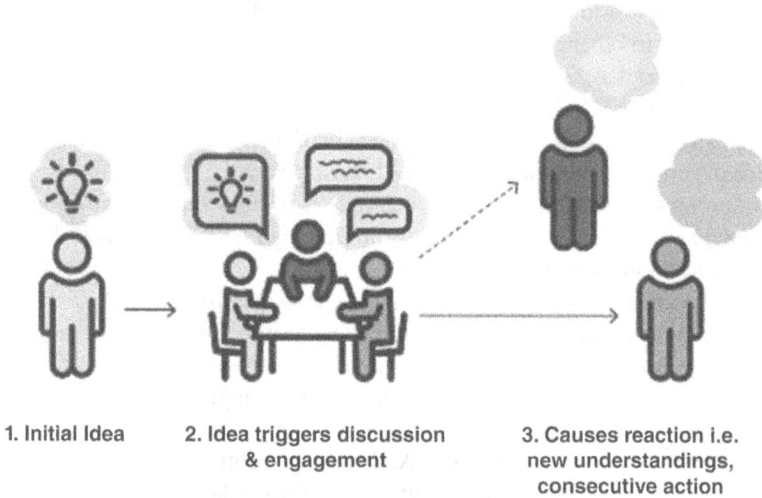

1. Initial Idea   2. Idea triggers discussion   3. Causes reaction i.e.
                              & engagement   new understandings,
                                                                 consecutive action

FIGURE 3.4   Summary of idea-as-trigger perspective.

idea development can have different functions, where actors try to gain access to information and new knowledge, enhance credibility, or exercise influence over to whom, where, and how the idea should proceed (Simon & Tellier, 2011). Support is essential not only to push ideas further and realize them, but also to sustain people's desire to do creative work for them (Baer, 2012; Deichmann et al., 2021). From this perspective, novel and unfeasible ideas may attract more collaboration and trigger learning (Deichmann et al., 2021) because they are perceived as interesting to work with, which could improve their chances of being implemented.

In summary, treating ideas as triggers-for-change helps capture, facilitate, and sustain the flow of ideas (Sosa, 2019) and see value in ideas not for their inherent properties, but in how the idea stimulates interpretation, discussion, and action (Sukhov et al., 2021; Gillier & Bayus, 2022; Deichmann et al., 2021). Idea development from an ideas-as-triggers-for-change perspective is about involving the right people at the right time, accessing and integrating new knowledge to a shared understanding, building legitimacy and support to sustain engagement, and pushing ideas towards realization (Baer, 2012; Mannucci & Perry-Smith, 2022). To foster innovation, it is necessary to create an environment that encourages the sharing of ideas, allows diverse interpretations, and nurtures engagement, discussion, and taking action. Ultimately, the power of an idea lies in its ability to act as a catalyst for change, setting in motion a series of interactions and engagements that can lead to meaningful and impactful outcomes (Figure 3.4).

3.4   CONCLUSION

Although stemming from the same definition that ideas are *provisional and communicable representations*, it is clear that there can be different interpretations and perspectives on what this entails in terms of idea development. Both ideas-as-objects and ideas-as-triggers perspectives represent an important function for innovation managers organizing idea development process and for the designers involved in idea development. Table 3.1 summarizes some of the most important aspects to consider when trying to improve idea development process.

Innovation managers and designers must balance treating ideas as objects and as triggers-for-change. Structured problem-solving, criteria-driven evaluation, and development of ideas can ensure clarity and feasibility but can stifle creativity. Mainly because structured approaches are seldom equipped to capture and

TABLE 3.1
Summary of the Idea Perspectives

Aspect	Ideas as Objects	Ideas as Triggers-for-Change
Concept	Ideas are distinct entities resulting from problem-solving activities	Ideas are enablers for actions, triggering new associations and discussions
Goal	To find, improve, and implement the best ideas	To realize value through engaging with ideas in various ways
Idea development	Focus on turning low-quality ideas into high-quality ones by elaborating, clarifying, and expanding	Emphasis on sustaining and directing interactions that arise during the design process
Evaluation criteria	Clarity, relevance, use value, workability, feasibility, specificity, and novelty	Ability to create discussions, draw attention, improve decision-making confidence, and gain legitimacy
Communication	Ideas need to be well-defined and easily understood	Ideas should create engagement and stimulate new interpretations and attachments
Ownership	Ideas can be owned and transferred legally and psychologically. Psychological ownership affects what kind of feedback is incorporated in the idea, while the potential for legal ownership (patents) can discourage people from sharing their ideas	Idea ownership is fluid and focuses on actions rather than legal rights. Psychological ownership is implicit and shared, focusing on stimulating action and interaction
Improvement focus	Enhancing idea quality by introducing new information and improving clarity	Improving idea sharing, activating social networks, and capturing engagement
Success factors	Traceability and identifying the best ideas through a survival-of-the-fittest approach	Capturing and sustaining the evolution of ideas, fostering diverse interpretations, and nurturing engagement
Main objective	Creative problem-solving to enhance idea quality	Facilitating environments where ideas trigger interactions leading to impactful outcomes

integrate new knowledge irrespective of the process stage. This can be seen during idea evaluation where experts infuse and generate new knowledge as part of their screening task, yet these generative actions are seldom made explicit (Boudier et al. 2023; Sukhov et al. 2025). Emphasizing ideas-as-triggers for engagement fosters innovative thinking but may lead to ambiguity. This is because the boundaries of what is the idea often becomes blurred in a discussion but remains alive in the minds of those involved. Managers should therefore create environments that encourage collaboration without excessive formalization. By introducing an idea-as-trigger perspective, an organization can help illuminate the creative work and efforts people contribute to innovation. This means that organizations should tailor their approaches not only to the needs of an efficient, structured innovation process, but also to the human needs and prerequisites that enable creativity. This can support designers in their solution-based, human-centric approach to solve ill-defined, 'wicked' problems (Li et al., 2022). Understanding the impact of ideas on people (Sukhov et al., 2019; Hua et al., 2022) can attract interest, spark engagement, and spur knowledge sharing – which is essential for good solution-based problem-solving and for sustaining creativity in organizations. Leveraging creative sparks (Boudier et al., 2023), encouraging discussions (Oak, 2011; Ter Wal et al., 2022), and guiding inspiration into actions (Deichmann & Jensen, 2018; Rouse, 2020; Mannucci & Perry-Smith, 2022) are therefore important elements that create a forward momentum in the design process.

Ideas as objects and as triggers-for-change are complementary, as each one provides valuable insights in understanding how idea development needs to be facilitated. The idea-as-objects perspective provides structure and clarity which guides designers in the quality and novelty of the ideas created and builds balanced ideation portfolios to drive different kinds of innovation projects (Kock et al., 2015). Ideas-as-objects perspective captures ideas in moments where they have a tangible form (Hua et al., 2022), i.e., a written description or sketch. In contrast, the idea-as-trigger perspective emphasizes action and engagement caused when people are exposed to ideas, initiating divergent thinking with regards to what the idea means or how they connect to other ideas. Ideas become alive as they are interpreted and infused with meaning. Ideas-as-triggers encapsulates more of the process as such which is characterized by divergent and convergent thinking, knowledge sharing and integration, legitimizing potential and getting support for next steps.

Turning low-quality ideas into high-quality ideas requires triggering refinement and expansion through discussions. Organizations that have embraced a structured approach to idea management might consider giving greater freedom for individuals to be inspired by the ideas they encounter, thereby nurturing a culture of sharing and long-term engagement for innovation. This is to avoid databases of unused ideas, since "ideas are useless if not used" (Levitt, 1963, p. 79). Designers who typically work with an iterative and collaborative problem-solving approach should be further encouraged to invite others into their process. They should view shared ideas not only as solutions but also as catalysts for engagement, support, and the cultivation of a creative culture. This approach can lead to better solutions and strengthen team dynamics.

In conclusion, a holistic understanding of both perspectives is crucial for improving ideas and sustaining the design process in organizations. Good ideas need support and engagement to reach realization.

REFERENCES

Amabile, T. M. (1982). Social psychology of creativity: A consensual assessment technique. *Journal of Personality and Social Psychology, 43*(5), 997–1013.

Amabile, T. M. (1998). How to kill creativity. *Harvard Busniess Review, 76*(5), 77–87.

Antons, D., & Piller, F. T. (2015). Opening the black box of "Not invented here": Attitudes, decision biases, and behavioral consequences. *Academy of Management Perspectives, 29*(2), 193–217.

Baer, M. (2012). Putting creativity to work: The implementation of creative ideas in organizations. *Academy of Management Journal, 55*(5), 1102–1119.

Baer, M. & Brown, G. (2012). Blind in one eye: How psychological ownership of ideas affects the types of suggestions people adopt. *Organizational Behavior and Human Decision Processes, 118*(1), 60–71.

Baruch, A., May, A., & Yu, D. (2016). The motivations, enablers and barriers for voluntary participation in an online crowdsourcing platform. *Computers in Human Behavior, 64*, 923–931.

Beretta, M. (2019). Idea selection in web-enabled ideation systems. *Journal of Product Innovation Management, 36*(1), 5–23.

Boudier, J., Le Masson, P., Netz, J., Sukhov, A., & Weil, B. (2023). Idea evaluation as a design process: Understanding how experts develop ideas and manage fixations. *Design Science, 9*(9), 1–22.

Brown, T. (2008). Design thinking. *Harvard Business Review, 86*(6), 84–92.

Campbell, D. T. (1960). Blind variation and selective retentions in creative thought as in other knowledge processes. *Psychological Review, 67*(6), 380–400.

Cohendet, P., Dupouët, O., Naggar, R., & Rampa, R. (2024). Orchestrating orphan ideas in the fuzzy front end of a large firm's R&D department. *R&d Management, 54*(3), 558–573.

Cooper, R. G. (1988). Predevelopment activities determine new product success. *Industrial Marketing Management, 17*(3), 237–247.

Cooper, R. G. (1990). New products: What distinguishes the winners? *Research-Technology Management, 33*(6), 27–31.

Cooper, R. G., & Kleinschmidt, E. J. (1987). New products: What separates winners from losers. *Journal of Product Innovation Management, 4*(3), 169–184.

Dane, E. (2010). Reconsidering the trade-off between expertise and flexibility: A cognitive entrenchment perspective. *Academy of Management Review, 35*(4), 579–603.

Dean, D. L., Hender, J. M., Rodgers, T. L., & Santanen, E. (2006). Identifying quality, novel, and creative ideas: Constructs and scales for idea evaluation. *Journal of the Association for Information Systems, 7*(10), 646–699.

Deichmann, D., Gillier, T., & Tonellato, M. (2021). Getting on board with new ideas: An analysis of idea commitments on a crowdsourcing platform. *Research Policy, 50*(9), 1–16.

Deichmann, D., & Jensen, M. (2018). I can do that alone… or not? How idea generators juggle between the pros and cons of teamwork. *Strategic Management Journal, 39*(2), 458–475.

Dorst, K., & Cross, N. (2001). Creativity in the design process: Co-evolution of problem–solution. *Design Studies, 22*(5), 425–437.

Eling, K., Langerak, F., & Griffin, A. (2015). The performance effects of combining rationality and intuition in making early new product idea evaluation decisions. *Creativity and Innovation Management, 24*(3), 464–477.

Florén, H., & Frishammar, J. (2012). From preliminary ideas to corroborated product definitions: Managing the front end of new product development. *California Management Review, 54*(4), 20–43.

Gillier, T., & Bayus, B. L. (2022). Group creativity in the wild: When building on ideas enhances the generation and selection of creative ideas. *Creativity and Innovation Management, 31*(3), 430–446.

Girotra, K., Terwiesch, C., & Ulrich, K. T. (2010). idea generation and the quality of the best idea. *Management Science, 56*(4), 591–605.

Hammedi, W., van Riel, A. C. R., & Sasovova, Z. (2011). Antecedents and consequences of reflexivity in new product idea screening. *Journal of Product Innovation Management, 28*(5), 662–679.

Hatchuel, A., & Weil, B. (2009). C-K design theory: An advanced formulation. *Research in Engineering Design, 19*(4), 181–192.

Hennessey, B. A. (1994). The consensual assessment technique: An examination of the relationship between ratings of product and process creativity. *Creativity Research Journal, 7*(2), 193–208.

Hua, M.-Y., Harvey, S., & Rietzschel, E. F. (2022). Unpacking "Ideas" in creative work: A multidisciplinary review. *Academy of Management Annals, 16*, 621–656.

Kahn, K. B., Evans Kay, S., Slotegraaf, R. J., & Uban, S. (Eds.). (2013). *The PDMA handbook of new product development* (3rd ed.). Wiley.

Karlsson, A., & Törlind, P. (2016). Mitigating lack of knowledge: A study of ideas in innovative projects. *International Journal of Design Creativity and Innovation, 4*(3-4), 144–161.

Kijkuit, B., & Van Den Ende, J. (2007). The organizational life of an idea: Integrating social network, creativity and decision-making perspectives. *Journal of Management Studies, 44*(6), 863–882.

Kock, A., Heising, W., & Gemuenden, H. G. (2015). How ideation portfolio management influences front-end success. *Journal of Product Innovation Management, 32*(4), 539–555. https://doi.org/10.1111/jpim.12217

Koen, P., Ajamian, G., Burkart, R., Clamen, A., Davidson, J., D'Amore, R., Elkins, C., Herald, K., Incorvia, M., Johnson, A., Karol, R., Seibert, R., Slavejkov, A., & Wagner, K. (2001). Providing clarity and a common language to the "Fuzzy front end. *Research-Technology Management, 44*(2), 46–55.

Levitt, T. (1963). Creativity is not enough. *Harvard Business Review, 41*(3), 72–83.

Li, W., Song, Z., & Suh, C. S. (2022). *Principles of innovative design thinking: Synergy of extenics with axiomatic design theory* (1 ed.). Springer.

Licuanan, B. F., Dailey, L. R., & Mumford, M. D. (2007). Idea evaluation: Error in evaluating highly original ideas. *The Journal of Creative Behavior, 41*(1), 1–27.

Liedtka, J. (2015). Perspective: Linking design thinking with innovation outcomes through cognitive bias reduction. *Journal of Product Innovation Management, 32*(6), 925–938.

Magnusson, P. R., Wästlund, E., & Netz, J. (2016). Exploring users' appropriateness as a proxy for experts when screening new product/service ideas. *Journal of Product Innovation Management, 33*(1), 4–18.

Maher, M. L., Poon, J., & Boulanger, S. (1996). Formalising design exploration as co-evolution. In J. S. Gero & F. Sudweeks (Eds.), *Advances in formal design methods for CAD* (pp. 3–30). Springer.

Mannucci, P. V., & Perry-Smith, J. E. (2022). "Who are you going to call?" Network activation in creative idea generation and elaboration. *Academy of Management Journal*, 65, 1192–1217.

Marcandella, E., Durand, M. G., Renaud, J., & Boly, V. (2009). Past projects memory: Knowledge capitalization from the early phases of innovative projects. *Concurrent Engineering Research and Applications*, *17*(3), 213–224.

Micheli, P., Wilner, S., Bhatti, J. S., Mura, S. H., & Beverland, M. (2019). Doing design thinking: Conceptual review, synthesis, and research agenda. *Journal of Product Innovation Management*, *36*(2), 124–148.

Moehrle, M. G. (2005). What is TRIZ? From conceptual basics to a framework for research. *Creativity and Innovation Management*, *14*(1), 3–13.

Naggar, R. (2015). The creativity canvas: A business model for knowledge and idea management. *Technology Innovation Management Review*, *5*(7), 50–58.

Nonaka, I. (1994). A dynamic theory of organizational knowledge creation. *Organization Science*, *5*(1), 14–37.

Oak, A. (2011). What can talk tell us about design?: Analyzing conversation to understand practice. *Design Studies*, *32*(3), 211–234.

Obstfeld, D. (2005). Social networks, the tertius iungens orientation, and involvement in innovation. *Administrative Science Quarterly*, *50*(1), 100–130.

Osborn, A. F. (1957). *Applied imagination: Principles and procedures of creative problem-solving* (3 ed.). Charles Scribner's Sons.

Osterwalder, A., & Pigneur, Y. (2010). *Business model generation: A handbook for visionaries, game changers, and challengers* (Vol. 1). John Wiley & Sons.

Perry-Smith, J. E., & Mannucci, P. V. (2017). From creativity to innovation: The social network drivers of the four phases of the idea journey. *Academy of Management Review*, *42*(1), 53–79.

Perry-Smith, J. E., & Shalley, C. E. (2003). The social side of creativity: A static and dynamic social network perspective. *Academy of Management Review*, *28*(1), 89–106.

Rouse, E. D. (2020). Where you end and I begin: Understanding intimate co-creation. *Academy of Management Review*, *45*(1), 181–204.

Seidel, V. P. (2007). Concept shifting and the radical product development process. *Journal of Product Innovation Management*, *24*(6), 522–533.

Seidel, V. P., & Fixson, S. K. (2013). Adopting design thinking in novice multidisciplinary teams: The application and limits of design methods and reflexive practices. *Journal of Product Innovation Management*, *30*(S1), 19–33.

Simon, F., & Tellier, A. (2011). How do actors shape social networks during the process of new product development? *European Management Journal*, *29*(5), 414–430.

Simonton, D. K. (1999). Creativity as blind variation and selective retention: Is the creative process Darwinian? *Psychological Inquiry*, *10*(4), 309–328.

Sosa, R. (2019). Accretion theory of ideation: Evaluation regimes for ideation stages. *Design Science*, *5*, 1–33.

Sukhov, A. (2018). The role of perceived comprehension in idea evaluation. *Creativity and Innovation Management*, *27*, 183–195.

Sukhov, A., Magnusson, P., & Netz, J. (2019). What is an idea for innovation? In P Kristensson, P. Magnusson, & L. Witell (Eds.), *Service innovation for sustainable business: Stimulating, realizing and capturing value from service innovation* (1 ed., pp. 29–48). World Scientific.

Sukhov, A., Sihvonen, A., Netz, J., Magnusson, P., & Olsson, L. E. (2021). How experts screen ideas: The complex interplay of intuition, analysis and sensemaking. *Journal of Product Innovation Management*, *38*(2), 248–270.

Sukhov, A., Sihvonen, A., Huck, J., Netz, J., & Olsson, L. E. (2025). How to manage generative idea screening. *Research Technology Management*, *68*(1), 35–45.

Ter Wal, A. L. J., Criscuolo, P., & Salter, A. (2022). Inside-out, outside-in, or all-in-one? The role of network sequencing in the elaboration of ideas. *Academy of Management Journal*, *66*(2), 432–461.

Tsoukas, H. (2009). A dialogical approach to the creation of new knowledge in organizations. *Organization Science*, *20*(6), 941–957.

Verganti, R., & Öberg (2013). Interpreting and envisioning: A hermeneutic framework to look at radical innovation of meanings. *Industrial Marketing Management*, *42*(1), 86–95.

4 Proposing a Design Thinking Framework to Develop Circular Products

J. A. Mesa and L. Ruiz-Pastor

By leveraging the human-centred design thinking (DT) process, the chapter introduces a five-step framework: discover, define, ideate, prototype, and test. This methodology incorporates tools like empathy mapping, life cycle analysis, and rapid prototyping to align user-centred insights with circular economy (CE) objectives. A detailed case study illustrates the framework's practical application in designing a sustainable disposable cup, showcasing strategies to reduce environmental impact and promote user engagement.

The integration of DT and CE fosters a holistic approach to product design, encouraging innovative solutions that are not only functional and desirable but also environmentally regenerative. This chapter aims to provide a formal guide for designers and stakeholders seeking to advance sustainable product development practices.

4.1 INTRODUCTION

The concept of the CE has gained relevance in recent years as a sustainable alternative to the traditional linear economy model. The CE emphasises the need to close the loop of product lifecycles through enhanced resource efficiency, waste reduction, and the continuous use of materials. By adopting this approach, CE aims to establish a regenerative system that benefits businesses, society, and the environment (Geissdoerfer et al., 2017; Kirchherr & Piscicelli, 2019).

Incorporating CE principles into product design involves rethinking how products are created, used, and disposed of. This approach includes strategies such as designing for durability, reparability, upgradability, and recyclability. It also involves considering the entire product lifecycle – from concept to end of life – with a focus on minimising environmental impact while maximising resource efficiency. Some key design principles include:

- *Design for longevity:* Products should be durable and easy to maintain and repair (Bocken et al., 2016).

DOI: 10.1201/9781003487524-4

- *Design for modularisation:* Products should be designed with inter-changeable parts to facilitate upgrades or replacements (den Hollander et al., 2017).
- *Design for disassembly:* Products should be such that they could be easily disassembled to promote recycling and reuse of components (Vanegas et al., 2018)
- *Material selection:* Sustainable, recyclable, or biodegradable materials should be prioritised (Walker et al., 2018)

Traditional product design often follows a linear, step-by-step process prioritising functionality, aesthetics, and cost-effectiveness. While this method successfully delivers innovative and marketable products, it often overlooks broader environmental and societal impacts. Furthermore, traditional design processes tend to be compartmentalised, with minimal collaboration across development stages, particularly concerning a product's end-of-life (Cross, 2021).

By contrast, DT is developed as a user-centred, iterative methodology that emphasises empathy, ideation, prototyping, and testing. This approach has revolutionised product development, fostering creativity and innovation by engaging a diverse range of stakeholders throughout the design process (Brown, 2008; Liedtka, 2015).

The integration of DT into the development of circular products represents a significant shift from the traditional producer-centric design paradigm. This integration promotes holistic thinking, encouraging designers to consider the entire system in which a product exists. By focusing on user needs and the broader context, DT helps identify opportunities to create products that are not only desirable and functional and desirable, but also sustainable and regenerative (Leifer & Steinert, 2011; Carlgren et al., 2016).

Combining the principles of CE with the methodologies of DT can lead to innovative solutions that tackle environmental challenges while addressing user needs. This approach involves understanding the environmental and social impacts of products from the user's perspective (Kolko, 2015; Brown, 2019), identifying opportunities for circularity in product lifecycles (Stickdorn et al., 2018), generating creative solutions that incorporate CE principles (Tschimmel, 2012; Ruiz-Pastor et al., 2022). Prototyping for longevity, modularity, and recyclability, followed by iterative refinement ensures that products are both sustainable and user-friendly (Curedale, 2016), and iteratively refining products based on feedback to ensure sustainability and user-friendliness (Martin & Hanington, 2012). Integrating these methodologies allows designers to create products that contribute to a more sustainable future, fostering innovation while considering environmental and societal impacts (Seidel & Fixson, 2013).

The goal of this chapter is to propose a DT framework specifically tailored for developing circular products. This framework integrates CE principles with user-centred, iterative methodologies of DT. The aim is to provide a structured approach that empowers designers to create products that are not only innovative and user-centric but also sustainable and regenerative. Through this framework, we will outline the steps and strategies involved and illustrate how they can be applied to real-world product development to address pressing environmental challenges and meet evolving consumer needs.

4.2 CIRCULAR ECONOMY

The CE concept is defined around three key principles: designing out waste and pollution, keeping products and materials in use, and regenerating natural systems. These principles drive a fundamental shift from a linear "take-make-dispose" model to a circular one, which aims to reduce resource consumption and environmental impact.

4.2.1 PRINCIPLES OF CIRCULAR ECONOMY

This principle emphasises the need to rethink products and processes to prevent waste at its source, moving from waste management to waste prevention (Tukker, 2015; Ellen MacArthur Foundation, 2021). Keeping products and materials in use focuses on extending their lifecycle through maintenance, reuse, refurbishment, remanufacturing, and recycling, thus reducing the need for new resources and minimising environmental impact (Geissdoerfer et al., 2017; Bakker et al., 2019). Regenerating natural systems aims to enhance the resilience of ecosystems and support their capacity to regenerate, ensuring that product design not only minimises harm but also contributes positively to environmental health (Korhonen et al., 2018; Stahel, 2016). Table 4.1 summarises the three principles of CE and their respective strategies regarding product design.

Incorporating these principles into product design promotes resource efficiency by reducing pressure on natural resources and minimising environmental damage amid growing global populations and consumption levels (Korhonen et al., 2018). It reduces waste by enabling products to be reused, refurbished, or recycled, thereby closing the loop and reducing pollution and landfill overflow associated with traditional linear models (Geissdoerfer et al., 2017). Furthermore, it offers economic benefits, such as opening new business opportunities, reducing costs related to raw materials and waste management, and enhancing supply chain resilience (Bocken et al., 2016). Finally, it helps companies meet the increasing consumer demand for sustainable, ethically produced goods, thereby improving brand reputation and fostering customer loyalty (Ghisellini et al., 2016); and it helps companies comply with stricter environmental regulations by reducing waste and emissions and promoting sustainable practices (Stahel, 2016).

4.2.2 CIRCULAR PRODUCT DESIGN

Circular product design integrates CE principles into the product development process, aiming to create products that are sustainable, resource-efficient, and capable of contributing positively to environmental regeneration throughout their lifecycle. Circular product design emphasises creating durable, modular, repairable, recyclable, non-toxic, and efficient products that minimise environmental impact while

TABLE 4.1

Summary of CE Principles and Strategies Related to Design and Lifecycle of Products

Principle	Strategy	Brief Definition	Authors
Designing out waste and pollution	Material efficiency	Reducing the amount of raw materials used in production	(Allwood et al., 2011; Tukker, 2015)
	Cleaner production	Implementing processes that reduce or eliminate the generation of hazardous substances	(Porter & Linde, 1995)
	Eco-design	Designing products that have minimal environmental impact throughout their lifecycle	(Brezet & van Hemel, 1997; Charter & Tischner, 2001)
Keeping products and materials in use	Product longevity	Designing durable products that last longer and can be easily repaired or upgraded	(Bakker et al., 2019; N. Bocken et al., 2019)
	Modular design	Creating products with interchangeable parts that can be replaced or upgraded	(Ellen MacArthur Foundation, 2013; den Hollander et al., 2017)
	Product-as-a-service	Shifting from selling products to offering them as services, retaining ownership and responsibility	(Mont, 2002; Tukker, 2015)
	Reverse logistics	Developing systems to take back products after use for reuse, remanufacturing, or recycling	(Blackburn et al., 2004; Guide & Van Wassenhove, 2009)
Regenerating natural systems	Eco-design	Creating products that minimise environmental impact by using sustainable materials and reducing energy consumption	(Charter & Tischner, 2001; Tukker, 2015)
	Closed-loop systems	Designing products and processes that allow materials to be continuously cycled back into production, reducing waste	(Stahel, 2016; Ellen MacArthur Foundation, 2021)
	Lifecycle assessment	Evaluating the environmental impacts of a product from cradle to grave to identify opportunities for improvement	(Finnveden et al., 2009; Guinée et al., 2011)

supporting the regeneration of natural systems (Mesa & González-Quiroga, 2023). This approach also requires a holistic consideration of the entire product lifecycle, from raw material extraction to end-of-life disposal, to maximise sustainability and economic benefits. Thus, circular products involve additional requirements in comparison to conventional products, which are usually designed to fulfil functionality, manufacturability, safety, and regulatory constraints (see Table 4.2).

With the additional requirements for circular products established, the following section examines how DT serves as a robust approach to address these challenges.

TABLE 4.2
Additional Requirements that Circular Products Need to Fulfil Compared to Conventional Products

Category	Requirement	Brief Definition
Sustainability	Renewable materials	Using materials that are renewable and sustainably sourced
	Recyclability	Designing for easy recycling at the end of the product's life
	Energy efficiency	Minimising energy consumption throughout the product lifecycle
Design for longevity	Durability	Ensuring the product lasts longer and withstands wear and tear
	Reparability	Making the product easy to repair by users or professionals
	Upgradeability	Designing the product to allow for upgrades to extend its useful life
Circular business models	Product-as-a-service	Offering products as services, retaining ownership and responsibility
	Reverse logistics	Implementing systems for taking back products for reuse, remanufacturing, or recycling
Design for disassembly	Modular design	Creating products with interchangeable parts that can be replaced or upgraded
	Easy disassembly	Ensuring products can be easily taken apart at the end of their life
Non-toxicity	Safe materials	Using materials that are non-toxic and safe for both users and the environment
Closed-loop systems	Material cycling	Designing products to allow materials to be continuously cycled back into production
Lifecycle assessment	Environmental impact	Evaluating and minimising the environmental impact of the product throughout its lifecycle
User engagement	User education	Engaging and educating users on maintenance, repair, and recycling

Source: Based on European Environment Agency (2017), Mestre & Cooper (2017), Zeeuw Van Der Laan & Aurisicchio (2019), Mesa (2023), and Mesa & González-Quiroga (2023).

4.3 DESIGN THINKING

4.3.1 ESSENTIALS OF DESIGN THINKING

In the context of CE, integrating DT emerges as a critical methodology for fostering innovation that addresses environmental challenges while meeting user needs. DT is a human-centred, iterative approach that drives the design and development of solutions by focusing on understanding users' needs and behaviours.

Its structured yet flexible nature allows for the exploration of complex problems, making it a valuable tool in the creation of circular products. The core stages of DT – empathise, define, ideate, prototype, and test – guide teams through this creative problem-solving process:

a. *Empathise:* This initial stage involves gaining a deep understanding of users' experiences, needs, and emotions. It is achieved through direct observation, interviews, and other user research methods. The goal is to develop an empathetic understanding of the users' challenges and behaviours, which becomes the foundation for the design process (Brown, 2008; Kolko, 2015).

b. *Define:* Insights gathered during the "empathise" stage are synthesised to create a clear problem definition. This stage involves translating user needs into a concise, focused problem statement that will direct the next steps of the design process. The Define stage sets the foundation for ideation by framing the right questions (Brown, 2008; Stickdorn et al., 2018).

c. *Ideate:* In this stage, diverse team members generate a broad range of ideas and potential solutions to the defined problem. The emphasis is on creativity, free thinking, and exploring a variety of perspectives. Brainstorming techniques are often employed to encourage the generation of novel and unconventional ideas without constraints (Brown, 2008; Tschimmel, 2012). At this stage, product concepts are developed, considering functionality, components, and possible improvements.

d. *Prototype:* Prototyping involves creating early, simplified versions of the potential solutions. These prototypes can be physical objects, digital interfaces, or process models. Prototyping enables the team to test ideas quickly, explore different possibilities, and gain valuable insights through experimentation (Martin & Hanington, 2012; Curedale, 2016).

e. *Test:* The final stage focuses on testing the prototypes with users to gather feedback. This iterative process allows designers to refine their solutions, ensuring they are practical, effective, and user-friendly. The Test stage often uncovers new insights, leading to further ideation and prototype development (Brown, 2008; Martin & Hanington, 2012).

4.3.2 Integrating Circular Economy and Design Thinking

Integrating DT with CE principles presents unique challenges and opportunities, driving innovation in the development of sustainable products and systems. One of the primary challenges is the complexity involved in addressing both user needs and environmental impacts simultaneously. This requires a deep understanding of systemic issues and is often difficult to achieve within traditional business models and processes (Geissdoerfer et al., 2017; Kirchherr & Piscicelli, 2019). Successful integration also demands effective stakeholder engagement, involving collaboration among designers, engineers, businesses, policymakers, and consumers. Aligning the diverse interests of these stakeholders and fostering effective communication can be a significant challenge (Kolko, 2015; Brown, 2019). In addition, implementing CE principles often entails higher initial costs for sustainable materials and processes, creating a resource constraint that must be balanced with the need for innovative, user-centred solutions (; Stahel, 2016; Korhonen et al., 2018).

Despite these challenges, there are substantial opportunities in combining DT with CE. This integration can lead to breakthrough innovations that address both users' needs and environmental sustainability (specially nowadays when users are more self-aware of eco-friendly solutions). By focusing on empathy and user insights, designers can develop products that are not only eco-friendly but also highly desirable (Brown, 2008; Tschimmel, 2012). DT also encourages the exploration of new sustainable business models, such as product-as-a-service and circular supply chains, which align with CE principles and can create new revenue streams and enhance brand value (Tukker, 2015; Ellen MacArthur Foundation, 2021). Furthermore, DT's emphasis on empathy and user involvement can foster greater consumer awareness and engagement in sustainability practices. Educated and engaged users are more likely to support and participate in circular initiatives, driving the adoption of sustainable practices (Stickdorn et al., 2018; Brown, 2019).

For successful integration of DT with CE, a holistic approach is essential. This involves embracing systems thinking, which considers the entire lifecycle of products and the interconnectedness of various stakeholders and processes (Geissdoerfer et al., 2017; Walker et al., 2018). Collaborative innovation is crucial, requiring cross-disciplinary collaboration and the involvement of all relevant stakeholders in the design process to ensure diverse perspectives and expertise are integrated (Kolko, 2015). Utilising iterative processes, such as prototyping and testing, allows for continuous refinement of solutions, ensuring they meet both user needs and sustainability goals. This agile approach provides flexibility and adaptation as new insights emerge (Martin & Hanington, 2012; Curedale, 2016). Finally, investing in education and awareness is vital. Educating designers, businesses, and consumers about the principles and benefits of CE and DT builds understanding and support for sustainable practices, fostering a culture of innovation and responsibility (Ghisellini et al., 2016).

4.4   THE PROPOSED FRAMEWORK

This framework integrates DT with CE principles to guide the development of sustainable products. It consists of five steps: discover, define, ideate, prototype, and test. And it is based on definitions provided by Carlgren (Carlgren et al., 2016), Brown (Brown, 2019), and Cross (Cross 2021). Figure 4.1 illustrates the structured flow of the proposed framework, moving through the stages of discover, define, ideate, prototype, and test. Each stage builds upon the insights and outcomes of the previous one, creating a cohesive process that drives both innovation and sustainability in product development. This flow emphasises a systematic approach to integrating CE principles, where each step contributes to refining the product's alignment with user needs and environmental objectives, ultimately ensuring a holistic and regenerative design process.

Iteration

1 Discover → **2 Define** → **3 Ideate** → **4 Prototype** → **5 Test**

Learn about user and circularity opportunities

Determine conventional and circular features

Brainstorm solutions based on circularity attributes

Simulate user experience

Validate with users

1.1 Empathy Mapping: gain understanding of users' needs, feelings and behaviors

1.2 LCA: understanding environmental impact of reference product across its lifecycle

2.1 Problem Statement: Define the problem and requirements

2.2 Definition of CE principles: Indentification of CE principles that can contribute to solve the problem

3.1 Mind Mapping: Brainstorming of ideas based con problem requirements

3.2 SCAMPER: Implement SCAMPER to booster initial ideas

4.1 Rapid Prototyping: Create quick and basic prototypes of potential solutions

4.2 Circularity Design validation: Implement design strategies (modularity, durability, reparability, upgradability)

5.1 User testing: User test the prototypes and generate feedback

5.2 Sustainability assessment: Evaluate the prototype's sustainability performance.

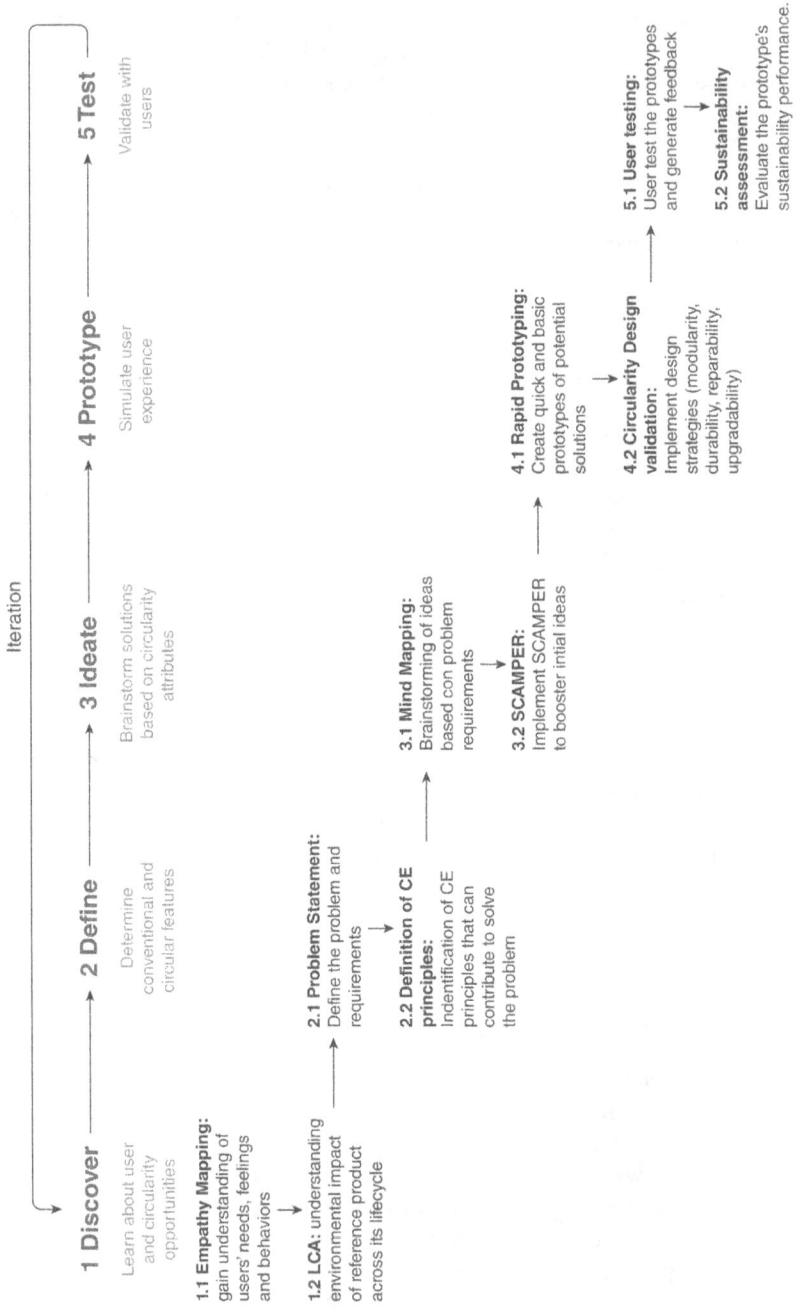

FIGURE 4.1 Proposed framework to generate circular products through design thinking.

4.4.1   DISCOVER: Empathy Mapping and Life Cycle Analysis

4.4.1.1   Empathy Mapping

The "discover" stage begins with Empathy Mapping, a tool that helps designers gain a deep understanding of users' needs, feelings, and behaviours (Bland, 2012; Thomas & McDonagh, 2013). This involves creating a visual map that captures what users say, think, do, and feel. By doing so, designers can empathise with their users, fostering a user-centred approach to design. Key points in this process include identifying pain points and desires, understanding user contexts and motivations, and using these insights to inform design decisions.

Life cycle analysis: Concurrent with empathy mapping, conducting a life cycle analysis (LCA) is essential for understanding the environmental impact of the product throughout its lifecycle. This includes material extraction, manufacturing, distribution, use, and end-of-life scenarios (Spreafico 2022). LCA helps identify hotspots for environmental impacts, guiding material selection and design choices to ensure the product meets sustainability criteria at every stage of its lifecycle.

4.4.2   DEFINE: Problem Statement and Circular Economy Principles

4.4.2.1   Problem Statement

After empathising with the users and conducting an LCA, the next step is to clearly define the problem you are solving. The problem statement should be a user-centric statement that focuses on their needs and how these relate to circularity (Brown et al., 2021). This step involves synthesising insights from empathy mapping and LCA to define a clear, concise, and actionable problem statement that highlights the connection between user pain points and environmental impacts.

4.4.2.2   Definition of CE Principles

Integrating CE principles into the problem definition is crucial. This involves incorporating principles previously listed in Table 4.1, such as designing out waste, keeping products and materials in use, and regenerating natural systems. These principles help guide the development of sustainable solutions and ensure that the design process addresses both users' needs and environmental impact. Aligning product features with CE goals is essential to meet the dual objectives of user satisfaction and sustainability.

4.4.3   IDEATE: Mind Mapping and SCAMPER Technique

4.4.3.1   Mind Mapping

The Ideate stage involves using brainstorming tools such as Mind Mapping to generate ideas visually (Davies, 2011). Mind Mapping allows designers to explore various aspects of the problem and potential solutions in a structured manner. This tool encourages divergent thinking and creative exploration, helping to organise ideas around central themes or concepts and identify connections and relationships between different ideas.

4.4.3.2   SCAMPER Technique

Another essential tool in the ideation stage is the SCAMPER Technique (Eberle, 1971), which prompts creative thinking and innovation. SCAMPER stands for "Substitute, Combine, Adapt, Modify, Putto another use, Eliminate, and Reverse". Applying SCAMPER helps challenge existing assumptions and conventional thinking, exploring multiple dimensions of potential solutions, and generating a wide range of innovative ideas for circular product design. For instance, using SCAMPER helps to generate ideas related to substituting non-renewable materials with sustainable alternatives, combine functionalities to reduce waste, or eliminate harmful processes. By applying it, designers can generate innovative solutions that enhance product longevity, recyclability, and resource efficiency, all of which are key components of CE.

4.4.4   PROTOTYPE: Rapid Prototyping and Circularity Design Validation

4.4.4.1   Rapid Prototyping

Rapid Prototyping is employed to create quick and rough prototypes of potential solutions. This approach allows for the visualisation and testing of ideas early in the development process. Rapid prototyping facilitates quick iteration and refinement of ideas based on feedback, minimising investment in detailed development until concepts are validated. This helps explore different design options and ensure that ideas are feasible and practical. Concerning CE, **Table 4.3** lists the main aspects that need to be incorporated towards more circular design prototypes.

TABLE 4.3

Main Aspects to Be Included in the Ideation and Prototyping Process Regarding CE

Material Selection	Geometry Definition	Structure Definition	9R Considerations
Biodegradable	Stress concentration	Modularity	Upgradeability
Recyclable	Assembly interfaces		Repurposability
Easy repairable	Material identification		Reusability
Highly durable			
Environmental resistance (chemical)			
Mechanical resistance			

Source: Mesa (2023); Mesa & González-Quiroga (2023).

4.4.4.2   Circularity Design Validation

Integrating circular design strategies into prototypes ensures alignment with CE principles. Strategies such as modularity, durability, reparability, and upgradability are incorporated into the design. This involves designing for longevity and ease of maintenance, using modular components to facilitate repairs and upgrades, and ensuring materials and components can be recycled or repurposed. These strategies help create products that are sustainable and adaptable to future needs.

This step refers specifically to assessing how well a product aligns with CE principles, such as recyclability, durability, modularity, reparability, and the ability to reduce environmental impact across the product lifecycle. While rapid prototyping focuses on quickly developing physical or digital models to test functionality, usability, and overall user experience, Circularity Design Validation ensures that the prototype also adheres to circular design principles. This step goes beyond general prototyping by evaluating aspects like material selection for sustainability, ease of disassembly for recycling, and the potential for future upgrades or repairs.

For example, in the development of a modular smartphone, rapid prototyping would involve creating a model to test how users interact with the modular design and its overall functionality. Circularity Design Validation, however, would involve evaluating whether the materials used are recyclable, ensuring that the phone can be easily disassembled for parts replacement or recycling, and confirming that the modular design extends the product's lifecycle by allowing easy upgrades.

4.4.5   TEST: User Testing and Sustainability Assessment

4.4.5.1   User Testing

The Test stage involves engaging users in testing the prototypes and gathering feedback (Wartika et al. 2023). User testing is crucial for evaluating the usability, desirability, and alignment of the product with user needs. Collecting qualitative and quantitative feedback from users helps improve the design and functionality of the product, ensuring it meets user expectations and requirements. This iterative process refines the product based on real user experiences.

4.4.5.2   Sustainability Assessment

In addition to user testing, conducting a Sustainability Assessment is essential to evaluate the prototype's sustainability performance. This involves assessing metrics such as energy efficiency, recyclability, circularity, and environmental impact using LCA. The sustainability assessment identifies areas for improvement to enhance the product's sustainability and ensures that the final product meets both user needs and environmental goals.

A compilation of possible techniques, tools and methods to use within all the stages in the framework developed are shown in Table 4.4. This table offers a practical guide for selecting appropriate tools at each stage of the framework, allowing designers flexibility in approach while staying aligned with CE principles. When choosing a specific tool or technique, designers should consider factors such as the tool's ability to promote resource efficiency, its support for

TABLE 4.4

Summary of Tools and Methods Applicable Within the Proposed CE and DT Framework

	Discover	Define	Ideate	Prototype	Test
Conventional DT tools	User interviews Stakeholder interviews Surveys Data analysis Metrics Competitors Focus groups Observation Clustering insights Context mapping Customer journey maps	Personas Empathy maps User journeys Storyboards User stories Problem statement Narratives Assumptions mapping Task analysis Jobs to be done Comparative analysis	Brainstorm sessions Mind maps Affinity maps Storyboard Card Sorting User journeys User flows Information architecture Service blueprints Business model canvas Crazy 8s Design principles SCAMPER	Paper prototypes Micro-interactions Detailed user flows Mock-ups Interactive prototypes Wireframes High fidelity design Design hands-offs Design documentation HTML/JS prototypes	Usability testing Shadowing A/B testing SUS surveys Heuristic Evaluation QA Analytics Performance testing Observations Desirability evaluations Metrics Eye tracking
Additional tools	Lifecycle analysis, Environmental impact assessment	Value mapping, SWOT analysis, CE strategies hierarchy	TRIZ (theory of inventive problem-solving)	Eco-design checklist, CAD generation	Sustainability assessment, environmental performance evaluation, scenario analysis

Tools and Methods

iterative testing, and its adaptability for engaging stakeholders in sustainability-focused innovation.

However, it is important to clarify that not all tools capture CE principles to the same extent. Designers should prioritise tools that emphasise resource efficiency, modularity, and stakeholder engagement in sustainability, which align more closely with CE goals.

4.5 CASE STUDY

The following conceptual case study demonstrates how the five-step framework combining CE and DT can be applied to design a sustainable disposable cup (see Figure 4.2), which are commonly made from plastic and paper, and present significant sustainability issues. The production of disposable cups often involves the use of non-renewable resources, such as petroleum for plastic and trees for paper. This process contributes to deforestation, greenhouse gas emissions, and

FIGURE 4.2 Disposable cup.

environmental degradation. Additionally, disposable cups are frequently lined with polyethylene, making them difficult to recycle and resulting in a substantial amount of waste ending up in landfills or polluting natural environments. The decomposition of these cups can take hundreds of years, during which time they release harmful chemicals into the soil and water.

4.5.1 DISCOVER: Empathy Mapping and Life Cycle Analysis

4.5.1.1 Empathy Mapping

Customers often mention that they need an affordable, convenient cup for beverages. While they are concerned about the environmental impact, they prioritise convenience. They frequently use disposable cups for coffee or other beverages on the go and feel guilty about using plastic cups but find them essential for their daily routine. Insights from empathy mapping reveal a conflict between convenience and environmental concern, highlighting the need for a sustainable solution that does not compromise ease of use.

Empathy Mapping was chosen to understand in deep the user's needs, behaviours, and emotions, which is a crucial step in identifying pain points related to sustainability and convenience. Among the tools listed in Table 4.4, this method was prioritised because of its ability to foster a user-centred approach, which is vital when designing sustainable products that need to balance both environmental impact and user practicality.

4.5.1.2   Life Cycle Analysis

Currently, disposable cups are made of plastic or non-recyclable paper, with high energy consumption and waste in production. These cups are single-use with low recycling rates, and the majority end up in landfills, with few being recycled. The LCA identifies key environmental hotspots, guiding material selection and design choices to ensure the product meets sustainability criteria at every stage of its lifecycle.

LCA was selected because it aligns with the principles of the CE by allowing us to assess sustainability across every phase of the product's lifecycle. This helps guide design decisions, such as material selection and manufacturing processes, ensuring that each aspect of the product minimises resource use, waste, and emissions. Unlike other tools that focus on specific aspects of design, LCA provides a holistic view of environmental impact, which is crucial for achieving circularity and long-term sustainability in product design.

4.5.2   DEFINE: Problem Statement and Circular Economy Principles

4.5.2.1   Problem Statement

Based on the empathy mapping and LCA findings, the following problem statement was formulated: Design an affordable, convenient disposable cup that minimises environmental impact and aligns with the lifestyle of environmentally conscious consumers.

4.5.2.2   Definition of Circular Economy Modifications

This involved incorporating principles such as designing out waste, keeping products and materials in use, and regenerating natural systems. The aim was to create a cup that is either biodegradable or easily recyclable, explore options for reusability or return systems, and use materials that are sustainably sourced or contribute positively to the environment. These principles guide the development of sustainable solutions, ensuring that the design process addresses both users' needs and environmental impact.

4.5.3   IDEATE: Mind Mapping and SCAMPER Technique

4.5.3.1   Mind Mapping

The brainstorming process considered materials such as bioplastics, bamboo fibre, and recycled paper. The design ideas included modular design for easy recycling and embedding seeds for biodegradation. The system ideas explored returnable cup systems and incentives for recycling. Mind mapping encouraged divergent thinking and creative exploration, helping to organise ideas around central themes and identify connections between different concepts.

Mind mapping was selected during the ideation phase to visualise and explore various sustainable material options and design strategies. This tool was preferred for its structured yet flexible approach, allowing for creative exploration of the product's lifecycle and circularity. Compared to other brainstorming techniques like Crazy 8s or Affinity Maps, mind mapping provided a more holistic view, aligning well with CE principles by helping to organise diverse ideas into coherent strategies.

4.5.3.2 SCAMPER Technique

Traditional plastics could be replaced with compostable materials, paired with a digital app for tracking and rewards. The design could be adapted to fit existing coffee machines and cup holders, and a QR code could be added for tracing the cup's lifecycle (See Figure 4.3). Non-recyclable components could be eliminated, and a returnable or exchange program could be implemented. SCAMPER ensured a comprehensive approach to sustainable design by challenging existing assumptions and conventional thinking. In addition, this tool provides a structured method to challenge conventional product DT and stimulate innovation through circular principles. In particular, SCAMPER's focus on Substitute, Combine, Adapt, Modify, Eliminate, and Reverse allowed us to reimagine

FIGURE 4.3   Disposable CUP with new material and QR.

material choices, the modularity of the product, and its lifecycle. This technique was more suited to the product's environmental goals compared to other ideation tools, as it directly addresses product reusability and recyclability in line with CE principles.

4.5.4 PROTOTYPE: RAPID PROTOTYPING AND CIRCULAR DESIGN STRATEGIES

4.5.4.1 Rapid Prototyping

Two prototypes were developed: a cup made from bamboo fibre (Figure 4.4), coated with a thin layer of biodegradable bioplastic, and a recycled paper cup with embedded seeds, designed to biodegrade and grow plants (Figure 4.5). Rapid prototyping allowed for the visualisation and testing of ideas early in the development process.

4.5.4.2 Circular Design Strategies

Cup and lid were designed to be easily separable for better recycling. The materials were chosen to be robust for the duration of use. Although reparability and upgradability were not applicable for a disposable cup, the modularity ensured

FIGURE 4.4   Bamboo fibre cup.

FIGURE 4.5   Embedded seeds cup.

that components could be separated for recycling (Figure 4.6). These strategies helped create products that are sustainable and adaptable to future needs.

4.5.5   TEST: USER TESTING AND SUSTAINABILITY ASSESSMENT

4.5.5.1   User Testing

Feedback on the feel and convenience of the cup and the willingness to participate in return/exchange programs is collected. Users' reactions to the concept of a cup that can biodegrade and grow plants are also assessed. The feedback is used to refine the design and improve the user acceptance.

4.5.5.2   Sustainability Assessment

The sustainability assessment identifies areas for improvement to enhance the product's sustainability. It ensures that the final product meets both user needs and environmental goals, aligning with CE principles. With all the information extracted from applying the framework, the circular redesign of the disposable cup can be fully developed.

In this case study, the selection of specific tools directly shapes the design outcomes, operationalising the framework in a way that prioritises CE principles.

FIGURE 4.6   Cup with new product architecture.

For instance, the use of LCA during the "discover" stage provided critical environmental insights that guided material selection, while the SCAMPER technique in the "ideate" stage fostered creative exploration of modular and reusable components. Each tool was chosen to enhance CE-specific outcomes, ensuring that the design aligns with both sustainability goals and user needs. This practical application of the framework demonstrates how targeted tool choices can drive circular product development and produce measurable impacts on design features, such as material recyclability and ease of disassembly.

4.6   CONCLUSION

Designing circular products is a step needed to foster sustainability and promote the correct use of technical and biological resources. DT techniques are a suitable tool to introduce all the features needed in a product to be more circular. In this chapter, a five-step framework was developed with the aim to introduce CE features in product design, through DT techniques. The framework covers the first design stages: from the investigation of the users' needs to the development and testing of a first prototype. By applying the framework to the disposable cup case, the design incorporates key sustainable aspects such as material recyclability, modularity for ease of disassembly, and minimised environmental impact. This practical application demonstrates how the framework enables designers to create or transform products to align with CE principles. While the framework offers substantial benefits by guiding product innovation through DT techniques, certain limitations, such as the complexity of material sourcing and the initial costs of sustainable production, may arise. Overall, the framework enhances circular and sustainable product design by integrating DT methodologies that support iterative improvements in both functionality and environmental performance

REFERENCES

Allwood, J. M., Ashby, M. F., Gutowski, T. G., & Worrell, E. (2011). Material efficiency: A white paper. *Resources, Conservation and Recycling, 55*(3), 362–381. https://doi. org/10.1016/j.resconrec.2010.11.002

Bakker, C., Hollander, M., Hinte, E., & Zijlstra, Y. (2019). *Products that last: Product design for circular business models.* BIS Publishers.

Blackburn, J. D., Guide, V., Souza, R., & Van Wassenhove, G. C., L. N. (2004). *Reverse supply chains for commercial returns* (2nd ed., Vol. 46). California Management Review.

Bland, D. (2012). Agile coaching tip: What is an empathy map. Available: http://www. bigvisible. com/2012/06/what-is-an-empathy-map.

Bocken, N., Strupeit, L., Whalen, K., & Nußholz, J. (2019). A review and evaluation of circular business model innovation tools. *Sustainability, 11*(8), 2210. https://doi. org/10.3390/su11082210

Bocken, N. M., de Pauw, P., Bakker, I., & van der Grinten, C. (2016). Product design and business model strategies for a circular economy. *Journal of Industrial and Production Engineering, 33*(5), 308–320. https://doi.org/10.1080/21681015.2016. 1172124

Brezet, H., & van Hemel, C. (1997). *Ecodesign: A promising approach to sustainable production and consumption.* UNEP.

Brown, T. (2008). Design thinking. *Harvard Business Review.*

Brown, T. (2019). *Change by design: How design thinking creates new alterantives for business and society.* Harper Business.

Brown, P., Baldassarre, B., Konietzko, J., Bocken, N., & Balkenende, R. (2021). A tool for collaborative circular proposition design. *Journal of Cleaner Production, 297,* 126354.

Carlgren, L., Rauth, I., & Elmquist, M. (2016). Framing design thinking: The concept in idea and enactment. *Creativity and Innovation Management, 25*(1), 38–57. https:// doi.org/10.1111/caim.12153

Charter, M., & Tischner, U. (2001). *Sustainable solutions developing products and services for the future* (1st ed.). Greenleaf Publishing.

Cross, N. (2021). *Design thinking: Understanding how designers think and work.* Bloomsbury Visual Arts.

Curedale, R. (2016). *Design thinking: Process and methods manual.* Design Community College Inc.

Davies, M. (2011). Concepting mapping, mind mapping and argument mapping: What are the differences and do they matter? *Higher Education, 62*(3), 279–301. https//doi. org/10.1007/s10734-010-9387-6 den Hollander, M. C., Bakker, C. A., & Hultink, E. J. (2017). Product design in a circular economy: Development of a typology of key concepts and terms. *Journal of Industrial Ecology, 21*(3), 517–525. https://doi.org/10.1111/jiec.12610

Eberle, R. F. (1971). *Scamper games for imagination development.* (Illustrated by de June Kern Weber). DOK Publishers.

Ellen MacArthur Foundation. (2013). *Towards the circular economy Vol. 1: an economic and business rationale for an accelerated transition.*

Ellen MacArthur Foundation. (2021). *What is a Circular Economy?* Https://Www.Ellenmacarthurfoundation.Org/Circular-Economy/What-Is-the-Circular-Economy.

European Environment Agency. (2017). Circular by design: Products in the circular economy. In *European Environment Agency* (Issue 6). https://doi.org/10.1002/14651858.CD004884.pub3

Finnveden, G., Hauschild, M. Z., Ekvall, T., Guinée, J., Heijungs, R., Hellweg, S., Koehler, A., Pennington, D., & Suh, S. (2009). Recent developments in life cycle assessment. *Journal of Environmental Management, 91*(1), 1–21. https://doi.org/10.1016/j.jenvman.2009.06.018

Geissdoerfer, M., Savaget, P., Bocken, N. M. P., & Hultink, E. J. (2017). The circular economy: A new sustainability paradigm? *Journal of Cleaner Production, 143*, 757–768. https://doi.org/10.1016/j.jclepro.2016.12.048

Ghisellini, P., Cialani, C., & Ulgiati, S. (2016). A review on circular economy: The expected transition to a balanced interplay of environmental and economic systems. *Journal of Cleaner Production, 114*, 11–32. https://doi.org/10.1016/j.jclepro.2015.09.007

Guide, V. D., & Van Wassenhove, R. (2009). OR FORUM—The evolution of closed-loop supply chain research. *Operations Research, 57*(1), 10–18. https://doi.org/10.1287/opre.1080.0628

Guinée, J. B., Heijungs, R., Huppes, G., Zamagni, A., Masoni, P., Buonamici, R., Ekvall, T., & Rydberg, T. (2011). Life cycle assessment: Past, present, and future. *Environmental Science & Technology, 45*(1), 90–96. https://doi.org/10.1021/es101316v

Kirchherr, J., & Piscicelli, L. (2019). Towards an education for the circular economy (ECE): Five teaching principles and a case study. *Resources, Conservation and Recycling, 150*(June), 104406. https://doi.org/10.1016/j.resconrec.2019.104406

Kolko, J. (2015). *Design thinking comes to age.* Harvard Business Review.

Korhonen, J., Honkasalo, A., & Seppälä, J. (2018). Circular economy: The concept and its limitations. *Ecological Economics, 143*(January), 37–46. https://doi.org/10.1016/j.ecolecon.2017.06.041

Leifer, L. J., & Steinert, M. (2011). Dancing with ambiguity: Causality behavior, design thinking, and triple-loop-learning. *Information Knowledge Systems Management, 10*(1–4), 151–173. https://doi.org/10.3233/IKS-2012-0191

Liedtka, J. (2015). Perspective: Linking design thinking with innovation outcomes through cognitive bias reduction. *Journal of Product Innovation Management, 32*(6), 925–938. https://doi.org/10.1111/jpim.12163

Martin, B., & Hanington, B. (2012). *Universal methods of design: 100 ways to research complex problems, develop innovative ideas, and design effective solutions.* Rockport Publishers.

Mesa, J. A. (2023). Design for circularity and durability: An integrated approach from DFX guidelines. *Research in Engineering Design, 34*(4), 443–460. https://doi.org/10.1007/s00163-023-00419-1

Mesa, J. A., & González-Quiroga, A. (2023). Development of a diagnostic tool for product circularity: A redesign approach. *Research in Engineering Design, 34*(4), 401–420. https://doi.org/10.1007/s00163-023-00415-5

Mestre, A., & Cooper, T. (2017). Circular product design. A multiple loops life cycle design approach for the circular economy. *Design Journal, 20*, S1620–S1635. https://doi.org/10.1080/14606925.2017.1352686

Mont, O. K. (2002). Clarifying the concept of product–service system. *Journal of Cleaner Production, 10*(3), 237–245. https://doi.org/10.1016/S0959-6526(01)00039-7

Porter, M., & Linde, E. (1995). Toward a new conception of the environment-competitiveness relationship. *Journal of Economic Perspectives, 9*(4), 97–118. https://doi.org/10.1257/jep.9.4.97

Ruiz-Pastor, L., Chulvi, V., Mulet, E., & Royo, M. (2022). A metric for evaluating novelty and circularity as a whole in conceptual design proposals. *Journal of Cleaner Production, 337*, 130495. https://doi.org/10.1016/j.jclepro.2022.130495

Seidel, V. P., & Fixson, S. K. (2013). Adopting design thinking in novice multidisciplinary teams: The application and limits of design methods and reflexive practices. *Journal of Product Innovation Management, 30*(S1), 19–33. https://doi.org/10.1111/jpim.12061

Spreafico, C. (2022). An analysis of design strategies for circular economy through life cycle assessment. *Environmental Monitoring and Assessment, 194*, 180. https://doi.org/10.1007/s10661-022-09803-1

Stahel, W. R. (2016). The circular economy. *Nature, 531*(7595), 435–438. https://doi.org/10.1038/531435a

Stickdorn, M., Hormess, M. E., Lawrence, A., & Schneider, J. (2018). *This is service design thinking*. O'Reilly Media.

Thomas, J., & McDonagh, D. (2013). Empathic design: Research strategies. *Australasian Medical Journal, 6*(1), 1–6.

Tschimmel, K. (2012). Design Thinking as an effective Toolkit for Innovation. *Proceedings of the XXIII ISPIM Conference: Action for Innovation, Innovating from Experience.*

Tukker, A. (2015). Product services for a resource-efficient and circular economy: A review. *Journal of Cleaner Production, 97*, 76–91. https://doi.org/10.1016/j.jclepro.2013.11.049

Vanegas, P., Peeters, J. R., Cattrysse, D., Tecchio, P., Ardente, F., Mathieux, F., Dewulf, W., & Duflou, J. R. (2018). Ease of disassembly of products to support circular economy strategies. *Resources, Conservation and Recycling, 135*(July 2017), 323–334. https://doi.org/10.1016/j.resconrec.2017.06.022

Walker, S., Coleman, N., Hodgson, P., & Collins, N. (2018). *Design for life: Creating meaning in a distracted world*. Routledge.

Wartika, Ulfah, A. P., Wahyuni, Melian, L., Hasti, N., & Alfariski, M. A. (2023). Website user interface design using the design thinking method. In 2023 International Conference on Informatics Engineering, Science & Technology (INCITEST) (Vol. 1, pp. 1–7). https://doi.org/10.1109/incitest59455.2023.10397000

Zeeuw Van Der Laan, A., & Aurisicchio, M. (2019). Designing product-service systems to close resource loops: Circular design guidelines. *Procedia CIRP, 80*, 631–636. https://doi.org/10.1016/j.procir.2019.01.079

5 The Startup Mindset

Innovation through Design Thinking

Karim Morcos

5.1 THE UNIVERSAL IMPORTANCE OF DESIGN THINKING IN TECH STARTUPS

In Chapter 1, the design thinking process is described as "never ending" and "a system of overlapping spaces". The iterative nature of the design thinking approach allows for constant learning and adaptation. Its focus on empathy ensures that solutions are deeply rooted in user needs, while experimentation and iteration foster flexibility, adaptability, and innovation. This approach leads to more effective and user-friendly products and services and encourages a more agile and responsive organizational culture. This is no different for tech startups. This chapter will delve into frequent missteps made by tech founders where the implementation of the design thinking process would have been crucial. It will also examine scenarios in which teams diligently applied design principles and subsequently reaped the benefits.

The design thinking approach is crucial across the foundational pillars of the tech industry – *software development, digital services, hardware, and tech-enabled services*. These sectors encompass a vast spectrum of business models, each integral to the technological landscape. By emphasizing a user-centric approach, design thinking ensures that innovations across these diverse areas are not only technically sophisticated but also meticulously aligned with user needs and preferences. This alignment is vital for the success of any technological product or service. The following sections delve into how design thinking fosters innovation through empathy within these essential sectors.

5.1.1 SOFTWARE DEVELOPMENT

In software development, design thinking helps teams to create more intuitive and user-friendly applications. By starting with empathy, developers can understand the frustrations, needs, and desires of their users, leading to software that solves real problems and provides a seamless user experience. This approach can lead to the development of features that users truly need, rather than what developers assume they need, thereby increasing user satisfaction and adoption rates.

 DOI: 10.1201/9781003487524-5

5.1.2 DIGITAL SERVICES

The term digital services specifically refers to services that are delivered online. This includes many services such as streaming media, cloud computing, and online education. Design thinking ensures that services are crafted around the user's journey, making complex processes simple and accessible. Through empathy, service designers can identify pain points in the service delivery process and innovate to enhance convenience, efficiency, and satisfaction. This user-first approach can transform a standard service into an exceptional one, fostering loyalty and differentiating the service in a crowded marketplace.

5.1.3 HARDWARE

In the hardware sector, design thinking goes beyond the technical specifications to consider the user's interaction with the device. Empathy allows designers to understand how, when, and why users interact with their hardware, leading to ergonomic designs, intuitive interfaces, and solutions that truly fit into the user's life and work. This can result in products that are not only functionally superior but also emotionally resonant with users, creating a stronger brand connection.

5.1.4 TECH-ENABLED SERVICES

These are traditional services enhanced by technology, such as fintech, health tech, or real estate tech. Design thinking bridges the gap between technology and human experience. By empathizing with users, companies can tailor their services to meet specific needs, making technology more accessible and useful to a broader audience. This can involve simplifying complex technologies, creating more personalized user experiences, or designing services that anticipate and solve user problems before they arise.

An example of how using design thinking can be used to enhance tech startup is a story of a company that was eager to transition from traditional paper-based workflows to a more modern digital framework. Their ambition was clear: to revolutionize their existing processes. Despite several strategy sessions with the operations manager (OM) and numerous iterations of mind maps to visualize the transition, there was a critical gap. It was suggested the inclusion of perspectives from the operational staff—those who engaged with these processes daily. The OM was confident that staff discussions had been sufficiently comprehensive to devise a robust digital solution; however, overlooked was a deeper understanding from various stakeholder viewpoints. The OM certainly had a profound grasp of the overarching processes, but it was the operational staff who illuminated the true intricacies and challenges inherent in the existing system. Their firsthand experiences revealed several pain points that had been overlooked. This discovery proved to be pivotal; it shifted the focus from merely digitizing an outdated system – with all its embedded flaws – to reimagining a solution that addressed critical issues while significantly enhancing the overall user experience.

This paradigm shift propelled beyond simple digitization, steering the company toward a transformative solution that was not merely a digital replica of the existing process, but a refined, user-centric system. By integrating direct insights from all levels of operation, a solution that truly resonated with the needs of those it was designed to serve was crafted, thereby not only meeting but exceeding the initial objectives for the project.

The team started off by digitalizing the manual workarounds, but two things happened that helped develop cross-team collaboration. First, the operations team understood what technology has to offer and started to ideate with this new "tool". Second, each team member felt their contribution was valued, which kept them more invested.

One notable enhancement involved streamlining the processes related to payment and service delivery. Previously, these processes were considered sacrosanct, and believed to be essential for ensuring client payments, especially since services were delivered over several weeks, and payments were often made post-delivery. The fear was that altering them might de-stabilize the business model. However, a cumbersome system was replaced – characterized by multiple signatures, several emails, and stacks of paperwork – with a smart contract. This contract automatically generates proof of service once payment is processed and service delivery is documented and communicated to clients. This shift simplified operations and removed unnecessary reassurances, focusing on what was essential for service execution.

By truly understanding the people we were designing for, we could create a product that not only meets users' current needs but also anticipates future desires, setting the stage for sustained innovation and growth. Startups can harness the power of design thinking to achieve a competitive edge and innovation by empathizing deeply with their users, gaining profound insights into user needs, and tailoring their solutions to meet these demands. This user-centric approach to product design not only allows startups to stand out in crowded markets but also ensures that their offerings are directly aligned with user expectations. Through the iterative processes of prototyping and testing, startups can make data-informed decisions, significantly mitigating the risk of product failure by validating concepts early and refining them based on real feedback. Moreover, design thinking encourages a culture of collaboration within startups, breaking down traditional functional silos and bringing together diverse perspectives. This collaborative environment fosters the development of holistic solutions that are innovative, user-focused, and more likely to succeed in meeting the complex needs of their target audience. However, startup founders occasionally overlook the clear benefits of design thinking, favoring their own biases over empirical evidence.

5.2   THE NEGLECT OF DESIGN THINKING

5.2.1   Technology-First Approach

Startups, often captivated by their technological genius, can fall into the trap of prioritizing product features over the user experience and underlying needs. This misalignment between what the product offers and what users need or want can lead to dissatisfaction and challenges in achieving a proper product-market fit.

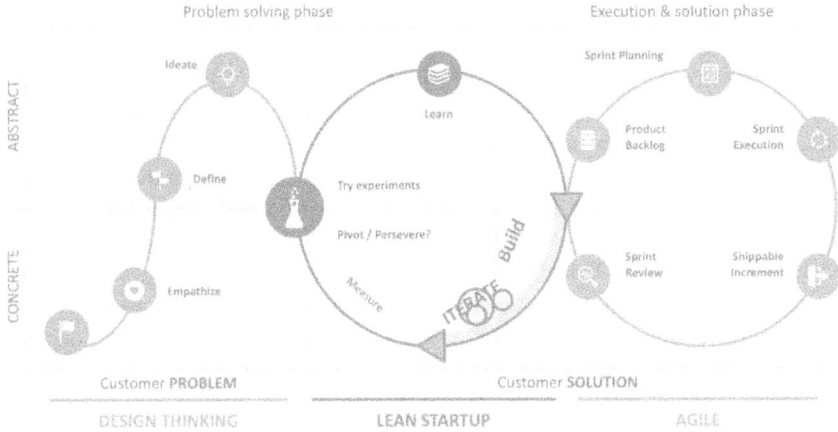

FIGURE 5.1 Overlapping of design thinking, lean startup methodology, and agile framework. (Image from PoweredTemplate.com).

Figure 5.1 shows how design thinking, lean startup, and agile methodologies are interconnected and where they overlap. Antonio Ghezzi (2019) indicated that the adoption of the lean startup methodology was rising among tech startups. However, since lean startup primarily jumps into the solution space, the problem space does not receive the attention it deserves, potentially overlooking critical user insights.

Design thinking fills the critical gap by emphasizing understanding and defining the problem space, which ensures that the solutions developed are deeply rooted in user needs. A notable example of this misalignment is Google Glass (Figure 5.2).

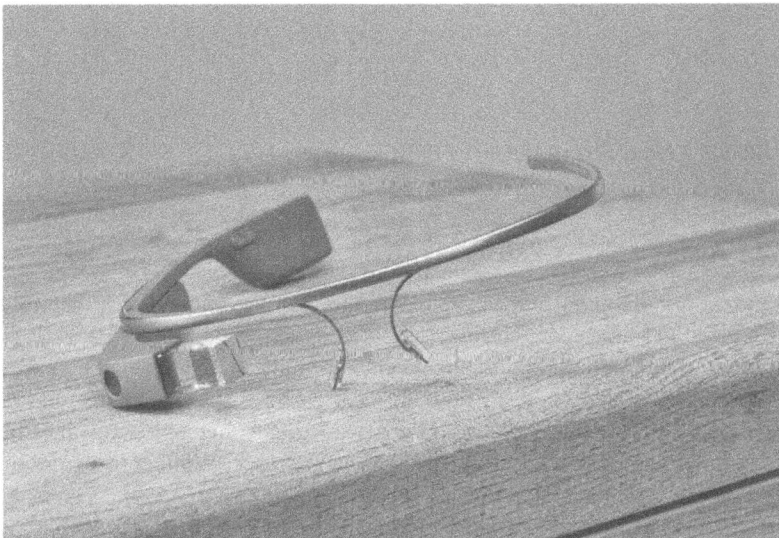

FIGURE 5.2 Google Glass. (Photo by Clint Patterson on Unsplash.)

The first edition was launched with much fanfare, and Google Glass was a technological marvel, showcasing the potential of wearable technology. However, it failed to resonate with the general public due to privacy concerns, a high price point, and a lack of clear, practical use cases for the average consumer. The focus was heavily on its technological capabilities without adequately addressing the real needs or concerns of potential users, leading to its decline as a consumer product.

Another example is the Juicero Press (Figure 5.3), a high-end juicer that was initially celebrated for its innovation. It was designed to squeeze proprietary, prepackaged juice packs, but consumers soon discovered that the packs could be squeezed by hand without the need for the expensive device. The product faced criticism for being over-engineered and not addressing any real consumer need that justified its cost, leading to the company's shutdown.

These examples highlight the critical importance for startups (or companies) to balance their excitement for technology with a grounded understanding of user needs and experiences. Focusing solely on technological capabilities without considering the practical value and relevance to users can result in products that, despite their innovative nature, fail to find a foothold in the market.

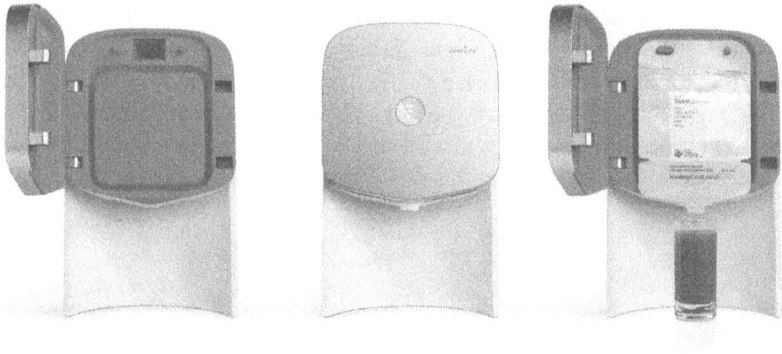

FIGURE 5.3   Juicero Press. (Photo by PR Handout.)

5.2.2 ASSUMPTIONS VS REALITY

A common pitfall among startups is the assumption by founders that they inherently understand their users' needs without conducting thorough research. This presumption can lead to the development of products that, while potentially innovative, fail to address real, pressing problems for the intended users. The foundation of a product's success lies not just in its technological advancement but also in its relevance and utility to its target audience. Without investing time and resources into understanding these users – their pain points, desires, and behaviors – startups risk creating solutions in search of problems, rather than addressing existing needs.

Equally detrimental is the lack of direct engagement with users. This is a step some startups skip due to time constraints, overconfidence, or simply undervaluing user feedback. Skipping user interviews, ignoring user feedback, or bypassing usability testing are shortcuts that can lead to a significant disconnect between what a product offers and what users want or need. This disconnect manifests itself in products that, despite their potential, fail to resonate with their target audience, leading to user dissatisfaction and product-market fit challenges.

This misalignment emphasizes the critical importance of embedding user research and feedback mechanisms throughout the product development process, ensuring that the end product is not just a reflection of the founders' vision, but a solution genuinely rooted in meeting user needs and expectations.

An example of the pitfalls of technology-driven development is illustrated by the author (see Figure 5.4). From day one, I was engrossed in coding an

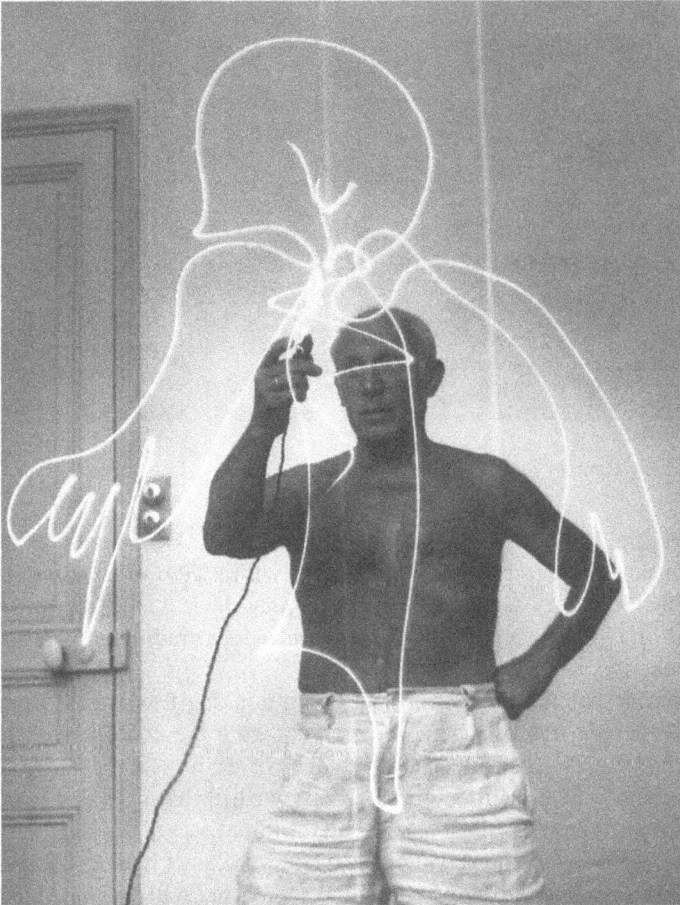

FIGURE 5.4   Lightpainting by Pablo Picasso (1949).

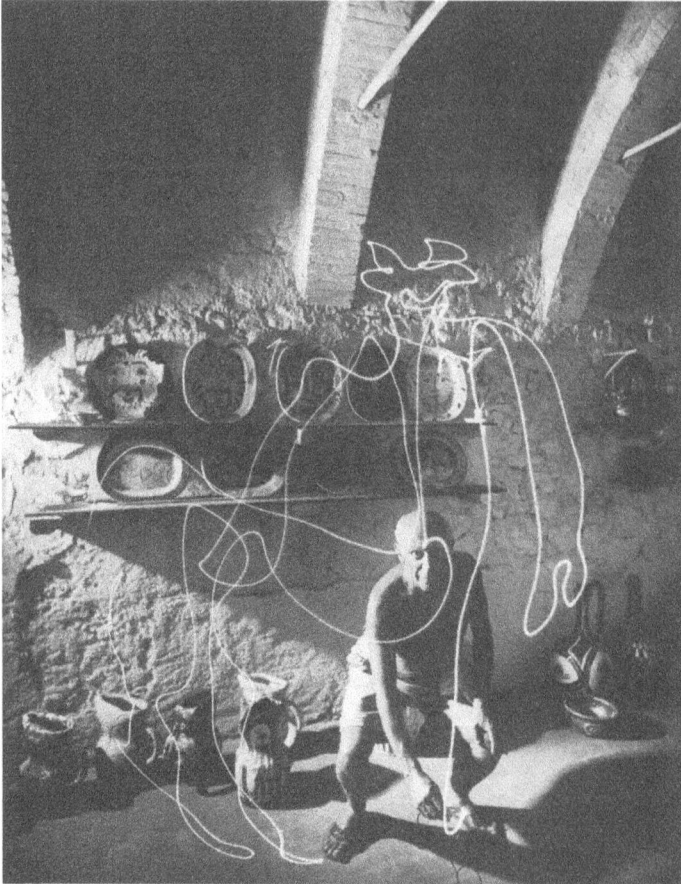

FIGURE 5.5   Centaur by Pablo Picasso (1949).

advanced app for video light painting, named PABLO in honor of Picasso's long-exposure photography (see Figure 5.5). Enthralled by the potential of integrating such complex technology into everyday devices, the aim was to simplify the process of creating long-exposure videos traditionally reserved for post-production.

Despite my late introduction to the design thinking framework, I proceeded hastily developing PABLO without first validating the concept with potential users. Initial user engagement consisted merely of sending surveys to professional photographers, which, while useful, barely scratched the surface of understanding user needs. The subsequent launch and the initial user uptake seemed promising with 10,000 registrations. However, the novelty quickly wore off, evidenced by a steady user churn. After revisiting the approach, PABLO 2.0 pivoted from a platform to a simpler filter app, but challenges persisted as user attrition continued. Examples of PABLO 2.0 are shown in Figures 5.6 and 5.7.

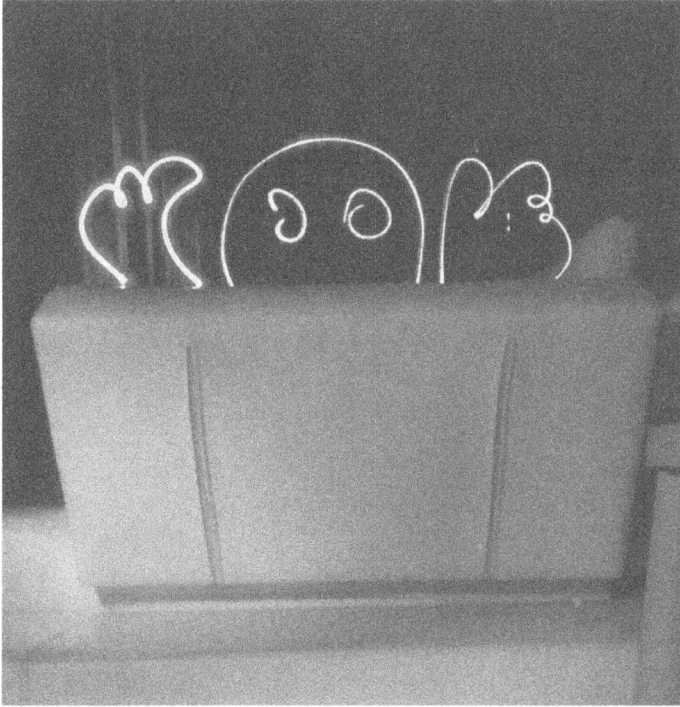

FIGURE 5.6   Lightpainting "Monster" created by the app PABLO.

Reflecting on this experience, it is clear that while we did engage in some phases of design thinking, such as ideation and testing, we significantly overlooked others. The empathy phase was notably overlooked; our user research was superficial and skewed towards a narrow user base – professional photographers – without diving deeper into the actual challenges & frustrations they faced. Moreover, the prototype phase was underutilized; we did not provide users with a prototype to interact with during the early stages of feedback. This oversight was compounded by assuming that technology alone was enough to attract and retain users, despite the observation that most of our user base did not consist of professional photographers. We later discovered that professional photographers often prefer using their sophisticated equipment and relish the manual process involved, an insight that was initially overlooked.

The journey with PABLO taught me the intricate dance of design thinking is not just checking boxes. It is about genuinely understanding and integrating user insights throughout every phase of product development. Now, as I develop a new SaaS product for developers, I am committed to involving a broader spectrum of stakeholders from the beginning. We are actively seeking to distinguish our assumptions from true insights through rigorous feedback loops. This engagement is crucial, particularly since our team includes members of our target audience,

FIGURE 5.7   Lightpainting "Carriage" created by the app PABLO.

emphasizing the need to continuously challenge our perspectives to shape a product that truly resonates with the broader community of developers.

In conclusion, embracing design thinking is an iterative and reflective practice that grows deeper with experience. It demands humility to acknowledge where we need to listen more and adapt faster, ensuring that our innovations not only solve real problems but also enrich the lives of those they are meant to serve.

5.2.3   CULTURAL AND ORGANIZATIONAL BARRIERS

It is crucial for organizations to create a fertile environment for harvesting innovation and overcoming cultural barriers (Hernández-Mogollon et al., 2010). The lack of a design advocate in a startup can lead to the undervaluation of design thinking in product development, as there is no dedicated leader to champion user-centered approaches and ensure these principles are embedded in the process. This oversight can skew development towards technical or business goals, neglecting user needs and the benefits of iterative, empathetic design practices.

Similarly, organizational silos – where engineering, design, and marketing teams work independently – hinder the collaborative essence of design thinking. Without cross-functional integration, it is challenging to achieve a unified vision that balances technical feasibility, aesthetic appeal, and market fit, ultimately impacting the product's success in meeting user expectations. Encouraging cross-disciplinary collaboration is crucial for leveraging diverse perspectives and fostering innovative, user-centric solutions. Unfortunately, organizational silos can become a self-perpetuating cycle, and once founders fall into these traps, they often defend and reinforce them with further psychological biases.

Sometimes, eagerness to launch hinders the existing design culture. An example of this, in Berlin, Germany, a tech startup, Imm-Tech, that specializes in immigration services for firms in Germany was looking to hire C-level executives from abroad, offering comprehensive support that extends beyond paperwork to include relocation assistance and school placements for families. Imm-Tech meticulously embarked on their journey with a deep dive into understanding their user base, employing desk research, interviews, and immersion techniques to map out user behavior. They pinpointed a specific challenge: facilitating the immigration process for families with young children, aiming to eliminate the logistical burden of childcare during this transition.

Despite their user-centered intentions, the startup faced a cultural roadblock when management pushed for the immediate launch of a product that had not yet been tested. This decision led to dissatisfaction among families, highlighting a critical oversight in the user journey – the lack of transparent communication about the process and additional costs for certain services. This scenario underscores the importance of visibility in the system's status to manage user expectations effectively.

Imm-Tech discovered that had there been prototype testing and focus group discussions, the team might have identified this gap earlier. Eventually, by incorporating user feedback, Imm-Tech developed a what-to-expect fact sheet, significantly improving customer satisfaction.

The case of Imm-Tech serves as a reminder that focusing solely on the functional/technical aspects of a product or service, while neglecting the importance of prototyping and user testing, can lead to solutions that fall short of expectations. Startups must balance their ambitions with thorough testing and user feedback to ensure their solutions truly meet the needs and enhance the experiences of their intended audience.

5.3   THE PSYCHOLOGICAL BIASES

David Hirshleifer (2015) defines behavioral finance as "the application of psychology to finance, with a focus on individual-level cognitive biases." Figure 5.8 shows that behavioral finance is the intersection of psychology, finance, and decision-making.

The origins of these biases vary from person to person, influenced by individual experiences. Everyone is susceptible to these biases, though the degree to which they are exhibited can differ significantly (Nobre et al., 2022).

Cognitive biases are systematic patterns of deviation from norm or rationality in judgment, where individuals create their own "subjective reality" from

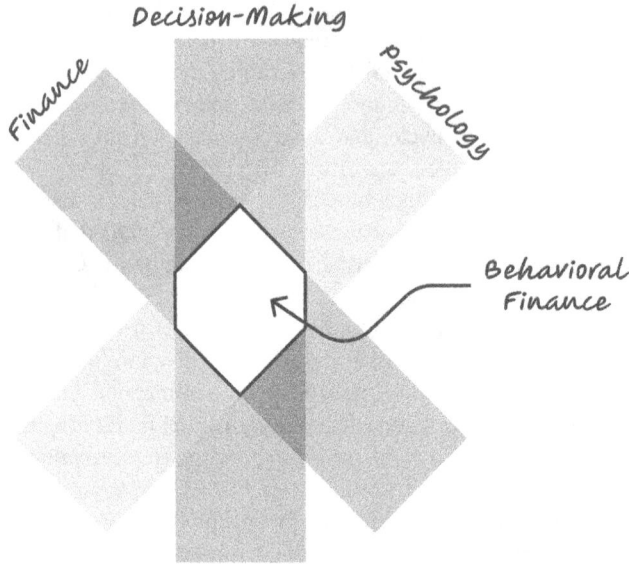

FIGURE 5.8 Behavioral finance: The intersection of psychology, finance, and decision-making.

their perception of the input. These biases significantly impact decision-making, often leading to errors in judgment, illogical interpretation, or irrationality in understanding others' actions and events (Madaan, G. & Singh, S., 2019). Startup founders are particularly susceptible to these biases due to the high-pressure and high-uncertainty environment that characterizes the startup culture. This setting often requires rapid decision-making with limited information, making founders more prone to rely on heuristic shortcuts and biases.

5.3.1 CONFIRMATION BIAS

Confirmation bias is when founders selectively seek out and give more weight to information that confirms their existing beliefs about their product or market, while disregarding or minimizing information that contradicts these beliefs. This bias can lead to a narrow understanding of the market and the true needs of users, causing startups to miss out on crucial opportunities for adaptation and improvement.

For example, the story of the BlackBerry by Research in Motion (RIM) serves as a poignant real-world example. In the early 2000s, BlackBerry dominated the smartphone market, particularly among business professionals, due to its innovative email service and physical keyboard. However, as consumer preferences began to shift towards smartphones with touchscreens and a broader range of apps, BlackBerry's leadership remained confident in their original product strategy (McNish & Silcoff, 2015). They continued to focus on their existing customer base and strengths, underestimating the appeal and potential of the emerging touchscreen smartphone market.

This confirmation bias led BlackBerry to dismiss the iPhone's launch in 2007 as a non-threat. The company's executives believed that their understanding of the market and their customers' needs was superior, especially those requiring secure email and a physical keyboard. As a result, BlackBerry delayed innovating their product line to include touchscreen technology and a more extensive app ecosystem. By the time BlackBerry recognized the shift in market demand and attempted to adapt, it had lost significant market share to Apple and Android devices, leading to its decline as a smartphone leader.

5.3.2 Overconfidence Bias

Overconfidence bias is where founders overestimate their own knowledge, abilities, and the potential success of their startup. This optimism can lead them to take unnecessary risks, underestimate challenges, or overlook crucial feedback from the market or their team. Founders with overconfidence bias may proceed with aggressive expansion, product development, or investment without adequately assessing the feasibility or demand.

A notable example of overconfidence bias is the case of Quibi, a short-form streaming platform launched by Jeffrey Katzenberg and Meg Whitman. Despite their extensive experience in entertainment and technology, the founders were confident that there was a significant market for premium short-form content, investing heavily in high-quality production and securing nearly $1.75 billion in funding before launch (Perez, S. 2022). They anticipated that millions of users would be willing to pay for content designed exclusively for mobile viewing, even with existing competition from YouTube, Netflix, and other free or established platforms.

However, Quibi failed to consider consumer feedback and underestimated the importance of content flexibility, such as the inability to share content on social media or watch on larger screens, which did not align with viewing habits (Julia Alexander, 2020). Furthermore, its launch in April 2020, amidst the global COVID-19 pandemic, when potential users were home and more likely to watch longer content on larger screens, exacerbated its challenges. Despite the founders' confidence, Quibi struggled to gain traction and announced its shutdown just six months after its launch.

Quibi's story illustrates how overconfidence in one's vision, without sufficiently accounting for market feedback and changing conditions, can lead founders to make decisions that jeopardize the startup's success. It underscores the importance of humility, market research, and adaptability in the entrepreneurial journey.

5.3.3 Sunk Cost Fallacy

Sunk cost fallacy is when individuals continue to invest time, money, or resources into a project, idea, or strategy due to the substantial investment they've already made, rather than based on the current and future value or viability of the project. This fallacy is driven by an emotional commitment to recoup spent resources, leading founders to throw good money after bad, rather than rationally cutting losses and redirecting efforts to more promising avenues.

A classic example is the story of Color Labs. Launched in 2011, Color Labs received an unprecedented $41 million in venture capital before its product – a photo-sharing app – was even released. Despite the massive investment and high expectations, the app failed to gain user traction upon launch (Ellis Hamburger, 2012). The substantial initial funding could have led the founders to persist with the project longer than rational analysis would suggest, attempting to justify and recoup the invested capital. The company pivoted several times, trying different strategies to make the app work, but ultimately, it could not overcome its initial shortcomings and lack of market fit. Color Labs was shut down in 2012 (Sarah Perez, 2022), serving as a stark reminder of how the sunk cost fallacy can cloud decision-making and lead to continued investment in failing ideas.

Another example is the story of Theranos, a health technology company that claimed to have revolutionized blood testing. Founded by Elizabeth Holmes, Theranos attracted substantial investment, totaling hundreds of millions of dollars, based on the promise of its technology. Despite growing evidence internally and from the medical community that the technology was not performing as claimed, the company continued to pursue and promote its solution. This persistence was driven, in part, by the sunk cost fallacy, as abandoning the project would mean admitting failure and losing the massive investment. The eventual exposure of technological shortcomings led to the company's dramatic downfall.

These examples highlight how the sunk cost fallacy can lead startups to continue down unviable paths. They emphasize the importance of evaluating projects on their current and future merits rather than past investments.

5.3.4 ANCHORING BIAS

Anchoring bias is when individuals rely too heavily on the first piece of information they receive (the "anchor") when making decisions. In the context of startups, this can mean that initial estimates, valuations, user feedback, or market research disproportionately influence founders' strategic decisions and adjustments. Even when new, more accurate information becomes available, the initial anchor can lead to inadequate adjustments, potentially steering the startup in the wrong direction.

A real-world example of anchoring bias in the startup world can be observed in the case of the initial public offering (IPO) of WeWork. WeWork's early valuations and the hype around its business model and growth prospects served as a strong anchor for the company, its investors, and the market. The co-working space startup was initially valued at an impressive $47 billion in early 2019, based on early financial projections and the charisma of its CEO, Adam Neumann (Reuters, 2023). This valuation set expectations for the company's performance and potential market impact (WIONews, 2023).

However, as more detailed financial information became available to the public and investors in the lead-up to its planned IPO, significant concerns were raised about the company's profitability, business model sustainability, and governance (WIONews, 2023). Despite this new, critical information, the initial high valuation and expectations acted as an anchor, leading to a delay in adequately adjusting the company's valuation and strategy (The Accountancy Cloud, 2023). It was only after intense scrutiny and backlash that WeWork's valuation was dramatically reduced, and its IPO was eventually pulled. The anchoring effect of its initially proposed valuation and growth narrative had a lasting impact on the company's strategy and public perception, contributing to a significant leadership and strategy overhaul.

This case exemplifies how anchoring bias, driven by initial valuations and expectations, can cloud judgment and lead to inadequate responses to new, vital information, underscoring the importance of remaining flexible and open to adjusting strategies based on evolving data and circumstances.

Interestingly, the same fallacies impact investors as well. Both entrepreneurs and investors are making similar commitments – entrepreneurs invest their time and effort into their startups, while investors put their money on the line. At its core, the decision-making process for both is fundamentally the same, shaped by elements of psychology and finance. Syed Aliya talks about how investors are impaired by the same biases.

5.3.5 Overcoming the Biases

Allison G. Butler and Roberto (2018) and Jelena Nikolic (2018) discussed how biases are interfering with innovation, and possible ways to overcome them. Recognizing and addressing cognitive biases within a startup team is essential for mitigating their impact on decision-making. The first step involves fostering awareness and education about these biases, both at an individual level and across the team, to ensure everyone understands how these biases can distort judgment and decision-making. In addition, seeking diverse perspectives is crucial in challenging entrenched biases and expanding the team's collective understanding. By incorporating a wide range of viewpoints into the decision-making process, startups can make more balanced and inclusive decisions, which are more likely to lead to successful outcomes.

Implementing structured decision-making processes is another key strategy for overcoming biases. By relying on data and analytics, startups can make decisions that are grounded in evidence rather than swayed by subjective opinions or emotional investments. This approach helps in making more rational and objective decisions. Furthermore, cultivating a culture that prioritizes feedback and adaptation encourages continuous learning and flexibility. Such a culture enables startups to respond swiftly to new information or changes in the market, making it easier to pivot strategies when necessary and ensuring that decisions are always aligned with the most current and relevant data.

These biases can manifest at any stage of the design process and the following includes benefits to the critical empathy phase and prototype and test phase for each bias.

5.3.6   CONFIRMATION BIAS

- *Empathy phase*: Engaging with a diverse user base to collect a broad spectrum of experiences and viewpoints challenges pre-existing beliefs.
  - *Benefits*: Expands the understanding of user needs, preventing tunnel vision and fostering more inclusive solution development.
- *Prototype and test phases*: Testing different prototypes across varied user groups provides concrete feedback that may support or contradict initial assumptions.
  - *Benefits*: Helps validate or refine ideas based on actual user responses rather than assumptions, leading to more user-centered products.

5.3.7   OVERCONFIDENCE BIAS

- *Empathy phase*: In-depth interactions with users reveal the complexity of their needs and challenges, which can temper founders' assumptions about their understanding of the market.
  - *Benefits*: Grounds decision-making in real-world complexities, reducing the risk of simplistic or overly optimistic projections.
- *Prototype and test phases:* Iterative prototyping and rigorous testing expose the product to real-world conditions, providing practical feedback on its effectiveness.
  - *Benefits*: Ensures that product development is responsive to actual user feedback, encouraging adjustments that are based on evidence rather than overconfidence.

5.3.8   SUNK COST FALLACY

- *Ideate phase*: Encouraging the exploration of multiple solutions helps prevent premature commitment to a single direction just because of previous investments.
  - *Benefits*: Promotes flexibility and creativity, allowing teams to explore the best possible solutions without bias towards past decisions.
- *Prototype and test phases*: Iterative testing and a readiness to pivot based on what works or doesn't work encourages decisions based on current and future value rather than past costs.
  - *Benefits*: Focuses resources on developing and refining solutions that show real promise, regardless of past investments.

5.3.9   ANCHORING BIAS

- *Ideate phase*: Generating a wide range of ideas and solutions helps dilute the impact of initial thoughts or data points that might otherwise dominate the development process.
  - *Benefits*: Encourages a more comprehensive evaluation of potential solutions, reducing the risk of missing better options.

- *Define phase*: Regularly revisiting and refining the problem statement based on ongoing research and feedback helps prevent the team from becoming fixated on initial perceptions.
  - *Benefits*: Ensures the problem definition and subsequent solutions evolve with a deeper understanding of user needs and market dynamics.

5.4   INNOVATION BEYOND TECH

The conventional view of innovation often focuses narrowly on technological breakthroughs new gadgets, software, or digital platforms that push the boundaries of what's technologically possible. However, this perspective overlooks the vast landscape of innovation that extends beyond just the technical realm. A broader definition of innovation encompasses not only technological advancements but also creative leaps in business models, process improvements, customer engagement strategies, and value proposition enhancements. These dimensions of innovation can dramatically alter market dynamics, often proving to be as transformative—if not more so—than purely technological innovations.

Several tech companies have achieved significant market success by prioritizing non-technical aspects of innovation. For example, Airbnb revolutionized the hospitality industry not through new technology but by introducing a novel business model that leveraged existing technology to connect travelers with hosts, thereby creating a new type of travel experience. Similarly, Uber transformed urban transportation by applying a new business model to the existing technology of smartphones and GPS, fundamentally changing how people book and share rides. Another notable example is Slack, which, while not the first messaging platform, reimagined team communication by enhancing user experience and integrating with numerous productivity tools, thereby carving out a significant niche in a crowded market. These companies exemplify how innovations in business models, customer engagement, and value propositions can lead to substantial market success, demonstrating that the scope of innovation extends far beyond the confines of technology alone.

5.4.1   BUSINESS MODEL INNOVATION

Innovative business strategies like subscription models, freemium plans, and platform-based ecosystems are reshaping industries by changing how companies deliver value. Subscription models have turned software purchases into ongoing services, exemplified by giants like Adobe and Microsoft, fostering stable revenues and lasting customer relationships. Freemium strategies, seen in Spotify and Dropbox, offer basic services for free while monetizing premium features, broadening user bases and enhancing revenue. Platform ecosystems, developed by Amazon and Apple, revolutionize retail and media by connecting independent sellers and creators with vast audiences.

Netflix's pivot from DVD rentals to online streaming exemplifies a successful business model transformation. Initially challenged by the advent of digital technology and evolving consumer demands, Netflix transitioned to a streaming model. Netflix not only navigated new challenges, licensing, streaming infrastructure, and influencing user behavior, but also redefined the entertainment landscape, showcasing the impact of innovative business models on achieving market success and industry disruption.

5.4.2 CUSTOMER EXPERIENCE AND ENGAGEMENT

Tech startups such as Airbnb and Slack have set themselves apart by prioritizing customer experience, transforming industries with intuitive and engaging platforms that foster user loyalty. Airbnb's user-centric design has revolutionized travel accommodation, making it more personal and trustworthy. Slack's seamless communication interface has improved workplace collaboration, creating a more enjoyable user experience. Meanwhile, Peloton has leveraged technology to build a vibrant community around its fitness platform, enhancing engagement through interactive and social features. These examples underscore how focusing on customer experience can be a key differentiator for startups in crowded markets.

An example of the complexity of innovation, and the need to focus on the user experience is the story of VRpeutic Inc. Founder "AK", with a robust foundation in Virtual Reality (VR) from his Ph.D. work on pain management, sought to transition from academic research to the commercial sphere. However, he soon confronted a pivotal challenge: while the technology was ready, its application – a bridge to product-market fit – remained elusive.

His return to Egypt marked the beginning of a transformative journey. Engaging with medical professionals, AK identified a rising need for innovative solutions in the realm of learning disabilities. This insight spurred the initial pivot of VRpeutic, necessitating a multidisciplinary team adept in research, development, and usability testing. An early prototype focusing on "Attention Skills" was developed, leveraging off-the-shelf VR headsets paired with simple software. Yet, testing this prototype with local medical centers unveiled critical gaps.

Observations and interviews with three key stakeholders – the medical staff, the patients' parents, and the patients themselves –revealed distinct challenges. Medical professionals needed guidance on using VR technology, parents were skeptical about introducing another screen-based intervention, and the primary users, the children, found the gaming-optimized headsets cumbersome and the controls overly complex.

Addressing these concerns, VRpeutic embarked on an iterative journey of prototyping and testing. While the weight of the VR headset posed a design dilemma, the team chose to focus on enhancing the software interface to simplify user interaction, an adaptation that mitigated the hardware's limitations without venturing into the complex hardware development space. Figure 5.9 shows one of the latest modules to date, aiding children with learning difficulties.

This story of VRpeutic is not just a tale of technological innovation but a testament to the power of design thinking. By embracing this framework, AK and

FIGURE 5.9 VRpeutic module aiding children with learning difficulties.

his team navigated through empathy, definition, ideation, prototyping, and testing phases, each step informed by direct feedback and keen observations. This approach not only fine-tuned their product to meet the nuanced needs of their users but also propelled the expansion of VRpeutic into new markets, including Canada and Vietnam.

AK's journey underscores how using design thinking to improve innovation and the user experience as an ongoing cycle of learning and adaptation, vital at every stage of a startup's development. Through empathy, prototyping, and testing, VRpeutic tailored its VR solutions to meet the nuanced needs of its users, showcasing the power of a user-centered approach in overcoming challenges and scaling new heights.

5.4.3 MARKET STRATEGY AND SEGMENTATION

Market segmentation and targeting niche markets have proven to be innovative strategies for startups, allowing them to find success by focusing on underserved or new segments. For example, Square revolutionized payments for small businesses by addressing their specific needs, which were largely ignored by traditional financial services. Similarly, Duolingo tapped into the vast potential of language learners by offering a free, gamified learning platform, standing out in the crowded education tech market. These cases demonstrate how flexibility, responsiveness to market feedback, and innovative market approaches can carve paths to success, even in well-trodden sectors.

5.4.4 COLLABORATIVE INNOVATION AND ECOSYSTEMS

Strategic partnerships and collaborations are crucial for startups, opening doors to new markets, technologies, and expertise. For example, the collaboration between

Airbnb and local tourism agencies has expanded their market reach and enriched their offerings, providing unique travel experiences. In addition, the partnership between Slack and software development tools like GitHub and Jira exemplifies ecosystem innovation, creating a more integrated workflow for users. Similarly, the collaboration between Uber and various city transit systems to integrate ride-sharing with public transportation showcases ecosystem innovation, enhancing value for urban commuters. These instances highlight how startups can significantly benefit from alliances, leveraging collective strengths to innovate and deliver enhanced customer value.

Innovation extends far beyond the realm of technology alone. It encompasses a holistic approach to rethinking how every aspect of a business operates – from its core value proposition and customer engagement strategies to its business model and market approach. Startups should embrace this expansive view of innovation, recognizing that sustainable competitive advantages often lie in the creative application of ideas across the entire business spectrum. By encouraging a culture of innovation that transcends technology, startups can uncover unique opportunities to differentiate themselves and achieve long-term success in an ever-evolving marketplace (Rivera-Vazquez, J. C., Ortiz-Fournier, L. V., & Rogelio, F., 2009).

5.5   RAPID PROTOTYPING AS A PATHWAY TO SUCCESS

The rich tapestry of Egyptian culture is significantly woven through its vibrant food traditions, where the art of dining extends far beyond mere sustenance to become a cornerstone of social interaction and communal life. This deep-seated food culture, interlaced with daily routines and familial bonds, sets the stage for a unique phenomenon: the universal challenge of deciding what to eat, humorously acknowledged in homes and workplaces alike as a daily deliberation.

In this context, El-Menus emerged in 2011, rooted in the heart of Cairo, initially as a repository of scanned restaurant menus. The founding team embarked on a grassroots journey, meticulously collecting and scanning paper menus from a broad spectrum of eateries, thereby laying the groundwork for an innovative platform. This initiative quickly garnered attention, attracting restaurants eager to feature their offerings on El-Menus, signaling a shift in how dining options were explored and selected.

As the platform gained momentum, it transformed into a comprehensive digital menu service. This evolution fostered a more interactive and dynamic relationship between restaurants and their patrons, with eateries actively managing their profiles and keeping their menus up to date. Gradually, El-Menus expanded further, venturing into the realm of food delivery, thereby enhancing its role in connecting people with the flavors of Egypt through a seamless digital experience.

The journey of El-Menus from a menu repository to a full-fledged food delivery service underscores how rapid prototyping can drive innovation, allowing startups to explore and realize their full potential by continuously adapting to and incorporating user feedback. Initially, the team's hands-on approach to gathering and digitizing menus was a form of prototyping that tested the viability of digitizing restaurant offerings. As the platform gained traction, it quickly adapted to user feedback, incorporating features like user ratings and restaurant-managed profiles, before ultimately expanding into food delivery.

5.5.1 SPEED, EFFICIENCY, & RISK MITIGATION

Chapter 1 talked about prototyping and how it helps evaluating the capabilities and limitations of your idea and identifying potential issues. Rapid prototyping significantly streamlines the development process, empowering teams to swiftly pinpoint and rectify issues, tailor their offerings to align with user preferences, and circumvent the dedication of resources to unfeasible features. This method champions a dynamic approach to product development, wherein prototypes serve as immediate, tangible representations of ideas for testing and validation. Through early and iterative testing, teams can quickly discern what resonates with users and what falls short, enabling a more focused and effective refinement of the product.

Moreover, rapid prototyping promotes the active involvement of users from the outset, fostering a development trajectory that is deeply intertwined with user expectations and requirements. By engaging users in the evaluation of prototypes, developers gain invaluable insights into user experiences, preferences, and pain points, ensuring the final product is not only functional but also highly attuned to the needs of its intended audience. This ongoing dialogue with users throughout the development cycle helps in sculpting products that truly resonate with the market, enhancing user satisfaction and loyalty.

Furthermore, rapid prototyping acts as a crucial risk management tool by facilitating the early testing of concepts and assumptions. This pre-emptive exploration allows teams to identify potential flaws or misalignments with market demands well before substantial resources are committed. By uncovering and addressing these issues early, rapid prototyping diminishes the risk of costly pivots or overhauls later in the development process, safeguarding against the expenditure of time and capital on ventures that are unlikely to succeed.

Rapid prototyping is a strategic approach that makes startups agile, user-focused, and resilient. It enables quick iteration, embraces user feedback, and facilitates early testing of concepts, significantly reducing development time and risk. This approach is not just about creating products; it's a philosophy that drives innovation, ensuring startups can quickly adapt to changes and meet user needs effectively. It's a key to staying competitive and responsive in fast-moving markets.

5.6 CONCLUSION

The design thinking approach is crucial and relevant across all stages of a start-up's journey, not just during the early phases. Its iterative and cyclical nature is at the heart of driving innovation. By continually revisiting the core principles of empathy, ideation, and experimentation, startups can ensure their products and services remain closely aligned with user needs and market demands. This ongoing process of learning, adapting, and evolving is what enables startups to navigate the complexities of the business landscape effectively, ensuring long-term success and relevance.

While common pitfalls and psychological biases, such as overconfidence, confirmation bias, and the sunk cost fallacy, can significantly derail startups, a design thinking approach serves as a countermeasure to these issues. It emphasizes empathy for the user, encourages diverse perspectives, and fosters a culture of prototyping and testing. This approach promotes open-mindedness and flexibility, allowing startups to challenge their assumptions and pivot based on feedback. By integrating design thinking into their strategy, startups can avoid becoming trapped by their biases, making more informed decisions that are grounded in real user needs and behaviors.

REFERENCES

Alexander, J. (2020). Quibi shuts down: Here's why the streaming service failed. [online] The Verge. Available at: https://www.theverge.com/2020/10/22/21528404/quibi-shut-down-cost-subscribers-content-tv-movies-katzenberg-whitman-tiktok-netflix. Accessed: February 2024.

Butler, A. G. & Roberto, M. A. (2018). When cognition interferes with innovation: Overcoming cognitive obstacles to design thinking: Design thinking can fail when cognitive obstacles interfere; appropriate cognitive countermeasures can help disarm the traps. *Research Technology Management*, *61*(4), 45–51. doi: 10.1080/08956308.2018.1471276.

Ghezzi, A. (2019). Digital startups and the adoption and implementation of lean startup approaches: Effectuation, bricolage and opportunity creation in practice. *Technological Forecasting and Social Change*, *146*, 945–960. https://doi.org/10.1016/j.techfore.2018.09.017.

Google Glass. (2020). https://unsplash.com/photos/white-usb-cable-on-brown-wooden-table-cXtJoazT6Bk?utm_content=creditShareLink&utm_medium=referral&utm_source=unsplash

Hernández-Mogollon, R., Cepeda-Carrión, G., Cegarra-Navarro, J. G., & Leal-Millán, A. (2010). The role of cultural barriers in the relationship between open-mindedness and organizational innovation. *Journal of Organizational Change Management*, *23*(4), 360–376. https://www.emerald.com/insight/content/doi/10.1108/09534811011055377/full/html

Hirshleifer, D. (2015). Behavioral finance. *Annual Review of Financial Economics*, *7*, 133–159. https://www.annualreviews.org/content/journals/10.1146/annurev-financial-092214-043752#html_fulltext

Ingraham, N. (2012). Color shuts down. *The Verge*. Available at: https://www.theverge.com/2012/10/17/3516428/color-shuts-down. Accessed: March 2024.

Kabbani, A. (2024). Interviewed by Karim Morcos. 27 February, Cairo.

Madaan, G. & Singh, S. (2019). An analysis of behavioral biases in investment decision-making. https://doi.org/10.5430/ijfr.v10n4p55

McNish, J. & Silcoff, S. (2015). *Losing the signal: The untold story behind the extraordinary rise and spectacular fall of BlackBerry* (1st ed.). Flatiron Books.

Nikolić, J. (2018). Biases in the decision-making process and possibilities of overcoming them. A Digital Archive of the University of Kragujevac. https://scidar.kg.ac.rs/handle/123456789/13325

Nobre, F. C., de Camargo Machado, M. J., Nobre. L. H. (2022). Behavioral biases and decision making in entrepreneurs and managers. *Revista de Administracao Contemporanea, 26*(suppl 1). https://doi.org/10.1590/1982-7849rac2022200369.en

Perez, S. (2022). CNN streaming service pulls a Quibi, will shut down a month after launch. TechCrunch. Available at: https://techcrunch.com/2022/04/21/cnn-streaming-service-pulls-a-quibi-will-shut-down-a-month-after-launch. Accessed: February 2024.

Reuters. (2023). Why WeWork failed and what is next. Reuters.com. Available at: https://www.reuters.com/business/why-wework-failed-what-is-next-2023-11-07/#:~:text=The%20company%20grappled%20with%20expensive,to%20stave%20off%20its%20bankruptcy. Accessed: March 2024.

Rivera-Vazquez, J. C., Ortiz-Fournier, L. V., & Rogelio, F. (2009). Overcoming cultural barriers for innovation and knowledge sharing. *FelixJournal of Knowledge Management, 13*(5), 257–270. https://www.emerald.com/insight/content/doi/10.1108/13673270910988097/full/html

Schewelow, V. (2024). Interviewed by Karim Morcos. 13 March, Berlin.

The Accountancy Cloud. (2023). WeWork's $2 billion disaster: What went wrong. Theaccountancycloud.com. Available at: https://theaccountancycloud.com/blogs/weworks-2-billion-disaster-what-went-wrong Accessed: March 2024.

WIONews. (2023). What caused the failure of WeWork and what will be the company's next move. WIONews.com. Available at: https://www.wionews.com/business-economy/what-caused-the-failure-of-wework-and-what-will-be-the-companys-next-move-656768. Accessed: March 2024.

6 Driving Change through Design Thinking
Shaping Innovation in GovTech

Marzia Mortati, Ilaria Mariani,
and Francesca Rizzo

6.1 INTRODUCTION: BACKGROUND AND DRIVING FORCES

Design thinking (DT) is largely recognised for having a transformative potential when applied in the public sector, as its integration has been proven to enhance public service innovation by fostering user-centricity, processes reorganisation, and organisational change (Junginger, 2013; Kimbell, 2015; Bason, 2017; Mortati, 2022; Mortati et al., 2022; Mortati et al., 2023). Several scholars have identified and discussed the principles and practices through which DT can help public organisations rethink their processes for public service design and implementation, while improving their collaboration with citizens and stakeholders (Yuan and Gascó-Hernández, 2019; Liu, 2021; Khan and Krishnan, 2021; Concilio et al., 2022; Mortati et al., 2023). In the contemporary scenario, the approach that DT underscores is increasingly relevant, as public organisations are required to reshape government-citizen interactions also due to the recent technological advancements and the European Union's ambitious goal of achieving full 100% online accessibility for key public services by 2030.[1] This shift thus implies introducing innovative approaches and reshaping organisational processes and structures to better integrate technology into public administration procedures as well as in public services. In this regard, the notion of GovTech has emerged as an acronym of Government Technologies, highlighting the integration of advanced technologies within government functions to improve service delivery, policy-making, and value creation (Mergel et al., 2019; Ubaldi et al., 2019). In this chapter, we introduce GovTech both as a theoretical notion and an area of practice that is seeking answers to the difficulties faced by public organisations in harnessing digital technologies. In describing this, we emphasise how GovTech strives to go beyond public-private partnerships integrating the notion of Digital Government by focusing on co-creation and citizen-centricity for the development of digital public services. This highly resonates with some of the recognised benefits of DT for public sector innovation while also the European Commission is emphasising

DOI: 10.1201/9781003487524-6

the relevance of DT and citizen engagement to achieve a sustainable European GovTech ecosystem.[2] However, scant literature has worked on connecting these two areas and guidance is missing on how DT can concretely support the enhancement of GovTech.

In our study, we analyse how DT principles and practices can be adopted and adapted to the GovTech context to foster its 'ethification' (van Dijk et al., 2021), indicating the inclusion of human values both in functional and non-functional requirements of public services (Bharosa, 2022). Our theoretical hypothesis is that the adoption of DT can be an effective lever to facilitate the establishment of the multiple helix innovation models needed for developing human-centric digital public services (Bharosa and Janssen, 2020). Given the nascent stage of research in this area, our primary research question is: *How can DT principles and practices be adopted and adapted by Governments to effectively support the development of GovTech solutions (human and technological requirements) as well as the establishment of the appropriate governance structures and institutional arrangements?*

We answer this question by adopting the four GovTech focus areas defined by Bharosa (2022) and analysing the relevant literature to examine how DT can provide benefits to each. We therefore analyse how DT can help enable the development of GovTech, ensuring services are not only technologically advanced but also citizen-centred, reflecting a shift from traditional bureaucratic processes to higher citizen-centricity.

The chapter mainly provides two key contributions: (i) the examination of existing literature showcasing the relevance of DT principles and practices for GovTech, and (ii) the proposal of a conceptual framework that reconnects DT principles and practices to the GovTech focus areas. The framework aims to provide clarity and guidance on how DT can contribute to shaping and improving GovTech in Europe, extending from how to embed creative and generative thinking to how to engage citizens and stakeholders in collaboratively developing solutions alongside governments. In this framework, we explain the reasons and benefits of DT for public sector innovation, highlighting how co-creation, experimentation, iteration, and organisational learning remain challenging to implement in governments. Furthermore, we stress how these processes require dedicated strategies to enable a more radical shift in both practices and mindsets. Thus, our conceptual framework bridges theoretical concepts with an operational perspective, analysing how DT can support GovTech in terms of governance structures, institutional arrangements, user, context understanding, and technological development.

The chapter concludes by discussing the relevance of organisational change and learning, as a means to a broader transformation that is necessary for achieving sustainable results beyond experimentation. Specifically, we explain the need to reconnect DT principles and practices with strategies for learning and organisational change to nurture a process of *reflection in action* (learning while doing) and *reflection on action* (learning after doing) (Argyris and Schön, 1978).

Iterating between these phases is one of the core ways in which public organisations can be supported in questioning and challenging established mindsets. The goal is a culture revision that can facilitate the embedding of new practices like DT and a broader transformation that can effectively help to shift towards new models and processes.

6.2   DESIGN THINKING FOR PUBLIC SECTOR INNOVATION

DT is now widely recognised as a holistic and impactful approach to innovation that significantly affects various organisational layers, including strategic, operational, and governance aspects (Elsbach and Stigliani, 2018). One of the first notable links between DT and public sector innovation was established in a 2013 report by the European Commission's Expert Group on public sector innovation. In that context, experts have emphasised the necessity for public organisations to create societal value through new or improved processes and services (European Commission – Directorate-General for Research and Innovation 2013), pointing out the need to move from a reactive approach to innovation (or the search for improvements when societal problems have already manifested), to one that is proactive, strategic, and systemic (or the proposal of improvements in public services that can – to some extent – anticipate societal problems). This call for an innovative approach looks both internally at organisational procedures and externally at service delivery, signalling a shift towards developing citizen-centred services through digital platforms, fostering a new entrepreneurial culture among public managers, and delivering more personalised and citizen-centred responses to public issues. This transformative direction was later confirmed by the 2017 Tallinn Declaration[3] (Tallinn Declaration 2017), which explicitly advocates for adopting user-centricity to improve public service design/delivery and revising existing public procedures.

Beyond these reports, the introduction of DT as an approach to public sector innovation has been discussed in the design literature as bringing multi-level implications. At the service level, DT can help rethink the user experience as well as the effectiveness and perceived quality of public services for the final users (Beckman and Barry, 2007; Brown 2009; Liedtka and Ogilvie, 2011; Liedtka 2015; Elsbach and Stigliani, 2018). At the organisational level, it can influence the internal culture of public employees by shifting mindsets and helping rethink the social contract between public institutions and society (Junginger 2013; Deserti and Rizzo, 2014; Mintrom and Thomas, 2018). Furthermore, DT can open new pathways for innovation and transform the processes through which public services are developed and delivered. Building on this broad influence, scholars have highlighted DT's role in enhancing several activities of public institutions from user engagement to the introduction of capacities aimed at bolstering public innovation through experimentation and learning (Beckman and Barry, 2007; Brown 2009; Liedtka and Ogilvie, 2011; Liedtka 2015; Elsbach and Stigliani, 2018). This includes, for instance, the ability to conduct small scale experiments and learning from user feedback or the possibility to steer ideation and implementation of innovations through iterative cycles of context analysis, ideation, experimentation, and learning.

Several scholars have recognised a few main principles underpinning DT for public innovation (see Table 6.1). These principles include the use of abductive reasoning, which complements induction and deduction, as the central logic in creative thinking. A strong orientation towards human-centricity enhances understanding the needs of end-users and the context in which a service is deployed. Engagement through co-design integrates citizens directly into the innovation process, allowing them to act as co-creators. Prototyping facilitates early assessment of solutions through rapid and iterative testing. Real-context experimentation enables solutions evaluation and further development by users in real-life settings helping to gauge their performance against expectations and situational applicability.

TABLE 6.1
The Main DT Principles

DT principle	Description	Reference
Abductive reasoning	Adoption of abductive reasoning as the main thought process to tackle challenges. This is distinguished from deductive and inductive logics and is further explained through the notions of 'lateral thinking' and 'framing and reframing', emphasising the identification of novel perspectives to tackle issues.	Dorst (2011); Drews (2009)
Human centricity and co-design	Innovation develops through an in-depth understanding of the needs of service users and their direct involvement as experts in the innovation process.	Brown (2009); Holloway (2009)
Centrality of prototyping	Experimentation is pivotal to learn-by-doing in iterative cycles. In DT, solutions are tested through 'quick and dirty' prototyping, facilitating early assessment.	Dorst and Cross (2001)
Experimentation in real contexts	Prototypes are used and assessed by end-users in real contexts to assess their features against expectations and contexts of use.	Evans and Terrey 2016; Trischler, Dietrich, and Rundle-Thiele 2019
Iterative and non-linear process	Early and fast prototyping through iterative development and testing cycles with end-users and other actors are fundamental for fast learning and consequent improvement of solutions.	Rizzo et al. (2018)
Experiential learning	Learning-by doing and learning-through-making are the preferred ways to explore problems and kick-start the innovation process	Deutschmann and Botts (2015); Kolb (1984a, b)

Source: Mortati et al. (2023).

Beyond recognising a few principles, the literature also underscores DT practices that can help enhance public innovation and rethink public service design and development. These contents are summarised in Table 6.2 (Mariani et al., 2023). They include the following: meaning creation is seen as a process capable of generating value by leveraging intangible benefits, such as symbolic and emotional relationships with products and services. Public formation is based on the

TABLE 6.2
The Main DT Practices Relevant to Public Services Innovation

DT practices	Description	Reference
Publics formation	It emphasises inclusivity by integrating diverse perspectives and making various voices heard, particularly through the incorporation of agonism to move away from consensual decision-making towards challenging mainstream positions. This practice is essential for creating a setting where multiple and often conflicting perspectives can be recognised and negotiated, allowing for more robust and democratic discourse.	DiSalvo (2010); Le Dantec and Christopher (2016); Venturini et al. (2015); Björgvinsson et al. (2012); Alshuwaikhat and Nkwenti (2002)
Meaning creation and sense-making	It involves a critical analysis of contexts, deep understanding of stakeholder needs, and the transformation of empirical data into meaningful insights that inform decision-making. It contributes to enhancing democratic participation, enabling public authorities to engage more effectively with citizens by providing a clear understanding of local needs and capturing factors influencing development of proposals.	Kimbell (2015); Grönroos and Voima (2013); Junginger (2014); Bogers et al. (2010); Dong and MacDonald (2017)
Co-production	It fosters significant collaboration between citizens and public authorities by co-developing digital public services and solutions, with citizens gaining a degree of decision-making power. This process not only harnesses the collective intelligence of local communities but also involves them as active participants in the process, enhancing the effectiveness and responsiveness of public services	Osborne et al. (2016); Seravalli et al. (2017); Bekkers et al. (2013); Durose and Richardson (2015); Blomkamp (2018)
Experimentation and prototyping	It bridges the gap between theoretical approaches and practical application, particularly in the realm of public services. Prototyping allows for the tangible exploration of new ideas through 'quick and dirty' prototypes that undergo early testing, thus enabling multiple iterations before finalising solutions. It helps adapt public services to real-world needs and contexts, as it involves citizens directly in testing and providing feedback on these solutions in controlled environments	McGann et al. (2018); Coughlan et al. (2007); Sanders and Stappers (2014); Tõnurist et al. (2017); Zimmerman et al. (2007)

Source: Mariani et al. (2023).

notion of infrastructuring, namely the process that identifies and forms social and material dependencies and commitments among those who constitute the public. Co-production suggests a different type of relationship between public service provider and user, where the second is not a passive consumer but an active producer of public value. Experimentation and prototyping are means to overcome barriers and constraints to implementation, capable of mitigating fear of failure and allowing the verification of hypotheses prior to large-scale roll outs.

Together, the DT principles and practices described (Tables 6.1 and 6.2) are relevant to tackle some of the setbacks that traditional approaches to digital transformation in the public sector frequently encounter, like the conventional provider-centric orientation. Common pitfalls include merely digitising existing offline processes without considering the impact on organisational structures, neglecting user needs, failing to integrate user experiences across public and private services, and overlooking the contexts in which digital services are accessed (i.e., diverse levels of digital literacy and accessibility, the request for more personalised services). Other barriers to DT implementation in the public sector, as recognised by Kimbell (2011, 2012) and Sangiorgi (2015), include issues such as scalability and sustainability of initiatives, risk aversion and bureaucratic resistance, professional culture and training challenges, and economic pressures. These barriers underscore the need for strategic approaches that not only foster innovation but also align with organisational capabilities and constraints. Overcoming these challenges necessitates a shift towards more responsive methods of designing and implementing digital public services (Szkuta et al., 2014; Mergel, 2016), emphasising principles of user-friendliness, security, accessibility, and efficiency as articulated in the Tallinn Declaration (Tallinn Declaration 2017). These are challenges that DT is promising to help tackle by centering on human needs and supporting public organisations in rethinking existing procedures.

6.3   INTRODUCING GOVTECH: DEFINITION AND CORE COMPONENTS

GovTech has emerged as a response to the challenges public authorities face in leveraging digital technologies to radically improve their service implementation and delivery, autonomously. Broadly speaking, GovTech involves the adoption by public organisations of innovative technological solutions often developed and operated by startups and small and medium enterprises (SMEs) (Filer, 2019). As described in the literature, the concept of GovTech underscores several key elements, including the importance of public-private partnerships to facilitate the formation of networks that include public agencies, companies, start-ups, and scale-ups, the uptake of a human perspective in the integration of technology in public services, and the creation of the appropriate governance structures and institutional arrangements. The combination of these elements aims to transform established processes towards social, economic, and environmental benefits as well as administrative efficiency and enhanced service quality.

The adoption of GovTech solutions has been significantly accelerated by the demand for remote service delivery brought by the pandemic, with an influx of recovery funds aimed at the digital transition, and innovative procurement models also facilitating the collaboration between public organisations and start-ups. This has put emphasis on GovTech solutions, with a surge of applications ranging from AI-enabled digital assistants to automated compliance systems and online judicial platforms. Although the discourse is vibrant, an accepted and shared definition of GovTech is missing. While academic literature on the subject is scarce, the European Commission and other bodies have contributed to an expanding range of grey literature that attempts at defining GovTech while exploring its barriers and implications (Kuziemski et al., 2022; Mergel et al., 2022). Despite this, the term suffers from conceptual ambiguity (Table 6.3), with some definitions overlapping with eGovernment and digital government because of the excessive focus on the role of technology and on the centrality of public-private partnerships (e.g. Klievink et al., 2016). However, GovTech also strives to go beyond these elements highlighting the need of innovations that are neither technology-push nor solely focused on the benefits that the collaboration with the private sector can provide.

As shown in Table 6.3, existing definitions stress the following core components of GovTech:

- *GovTech as a strategy for public sector transformation*: From this viewpoint, any effort by the public sector to enhance its operations through technology can be classified as 'GovTech'. This viewpoint on GovTech is clear and comprehensive. However, it is also too broad, encompassing many methods that aren't necessarily new while lacking a clear differentiation from related concepts like digital transformation.
- *GovTech as a result of highly transformative technological solutions*: Emphasising the technological products highlights innovation as a crucial aspect of GovTech. This includes innovation in business and service models as well. However, this perspective risks downplaying the public sector's role in GovTech, potentially undermining the public transformation component.
- *GovTech as the implementation or utilisation of tech solutions specifically designed for the public sector*: Focusing on the implementation of tech tailored to the public sector highlights the role of procurement. However, defining GovTech as the act of incorporating tech solutions is essentially tautological.
- *GovTech as a catalyst for advancing innovative startups*: Another key aspect is looking beyond public administration. The growth of GovTech creates beneficial development cycles beyond a few established players but also fosters development opportunities for new, innovative entrepreneurs. However, adopting the sole perspective of startups and other innovation players loses the necessary focus on the internal processes of public organisations.

TABLE 6.3

Dominant GovTech Definitions in the Academic and Grey Literature

	Typology	Description	Source
1	Scientific literature	GovTech refers to socio-technical solutions – that are developed and operated by private organisations – intertwined with public sector components for facilitating processes in the public sector.	Bharosa (2022)
2	Grey literature	The term GovTech refers to the use of emerging technologies and digital products and services by government from start-ups and SMEs – instead of relying on large system integrators. There are many – oftentimes competing – definitions of the term GovTech. Despite this diversity, most definitions share the following three common elements: the public sector engages with start-ups and SMEs to procure innovative technology solutions, for the provision of tech-based products and services, in order to innovate and improve public services.	JRC, Mergel et al. (2022)
3	Grey literature	GovTech is a whole-of-government approach to public sector modernization that promotes simple, efficient, and transparent government, with citizens at the centre of reforms.	The World Bank, Dener et al. (2021)
4	Grey literature	GovTech is understood as the ecosystem in which governments collaborate with an innovative private sector, particularly startups with experience in data, digital technology and pioneering methodologies; in order to solve public problems. As such, GovTech is an essential tool which accelerates public innovation, improving policies and public services. Moreover, GovTech also functions as a strategy for economic development, leading to a redirection of public spending into technology and towards local digital startups and small to medium businesses; which possess a higher added value and aid in boosting productivity.	CAF, Zapata (2020)
5	Grey literature	GovTech is an emergent innovation ecosystem in which private sector start-ups and innovative small and medium enterprises (SMEs) deliver technological products and services, often using new and emerging technologies, to public sector clients. Many GovTech companies work on challenges presented by emergent policy areas, or on problems where no solution was previously imagined as technically possible. The priorities of the GovTech ecosystem include improved efficiency and greater accountability in the public sector and its interactions with citizens. Building trust across the diverse stakeholders in the ecosystem is crucial for developing a thriving GovTech industry to serve the domestic public sector and to contribute to national economic growth.	Filer (2019)

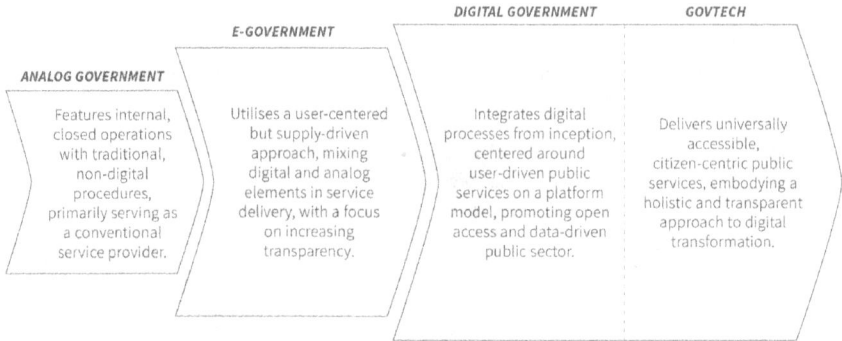

FIGURE 6.1 Elaborated from the phases of government digital transformation according to World Bank (2020, p. 4).

Further to this, the goal of GovTech is to drive a wider societal transformation, necessitating a socio-technical approach to its definition (Bostrom and Heinen, 1977; Mumford, 2006; Mortati et al., 2023). This perspective involves understanding social structures, roles, and rights to guide the design of systems that integrate communities of people and technology. By adopting this approach, GovTech can be seen as the network of people and organisations interacting through technological and institutional frameworks, altering behaviours from both technical and social standpoints. This definition helps highlight the difference between GovTech and other related notions. For instance, as shown in Figure 6.1, GovTech and digital government, while related, represent different stages in the integration of technology within the public sector. Digital government primarily focuses on re-engineering government processes to be digital by design, emphasising user-driven public services that are intuitive and responsive to citizen needs. It embraces the government-as-a-platform model, fostering an ecosystem where public and private entities can collaborate. Key characteristics of digital government include open-by-default policies that promote transparency and data sharing, a data-driven approach to public sector decision-making, and proactive administration that anticipates and addresses citizen needs pre-emptively. GovTech, as an evolution of digital government, advances this foundation by developing and deploying more sophisticated, often technology-driven solutions aimed at creating universally accessible, citizen-centric public services. It encourages a whole-of-government approach to digital transformation, striving to build simple, efficient, and transparent government systems. As such, GovTech also emphasises the role of startups and private sector players in driving public sector innovation, pushing beyond the digitalisation of existing processes to fundamentally transform service delivery and government operations.

6.4 METHODOLOGICAL APPROACH

Despite the increasing number of initiatives recently launched in Europe (Public, 2021; Kuziemski et al., 2022), GovTech is still largely in a nascent phase. Several challenges still hinder its development. Academically, GovTech lacks a unified framework and definition (Bharosa, 2022), resulting in a limited yet growing body of literature that supports its conceptualisation. Practically, several barriers hinder smooth adoption and implementation such as limited digital literacy and technological expertise within public institutions, outdated public procurement systems, and challenges in managing public-private relationships, which often lead to an overreliance on a few 'champions' of digitisation. Other significant challenges include the dependency on private sector technologies, which risks undermining public control and digital sovereignty, as governments may become reliant on commercial entities for critical services and infrastructure. In addition, there is a general lack of institutional frameworks that can effectively support GovTech initiatives, often leading to gaps in accountability and misalignment with public values. Finally, organisational cultures resistant to innovation pose significant barriers to embracing GovTech solutions. These challenges underscore the need for a strategic approach to GovTech that includes robust governance frameworks and careful consideration of the implications on public sector autonomy and citizen privacy. This approach involves operating at various complementary levels to facilitate systemic change and includes strategies that develop both inward (fostering an innovation culture) and outward (influencing market change). Specific strategies should be introduced to align design activities with strategic management and organisational objectives, enhancing design impact on core goals (Bason, 2017). By securing leadership buy-in and fostering cross-departmental collaboration, the implementation of a design strategy not only leverages DT as a transformative practice within the public sector but also reconciles public services conception and implementation. To elaborate on these

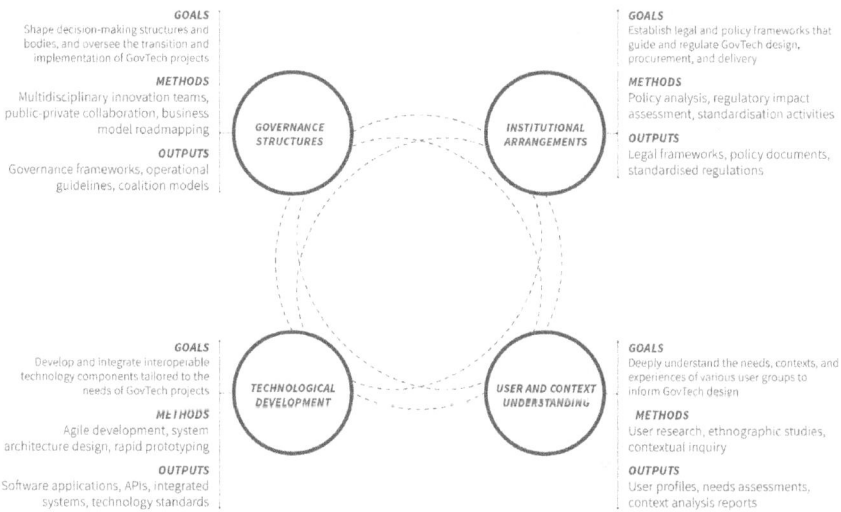

FIGURE 6.2 Adaptation of the conceptual framework for the GovTech design and governance of Bharosa (2022).

challenges and describe the benefits of DT adoption, we build on and adapt – for the purposes of this study – the socio-technical framework proposed by Bharosa (2022) (Figure 6.2), which identifies four main areas that need to be reinforced to support the effective design and governance of GovTech solutions: (1) governance structures, pertaining to adapting decision-making structures to allow co-creation; (2) institutional arrangements, looking at the necessary arrangements to legitimise GovTech procurement and delivery; (3) user and context understanding, highlighting the necessity to co-create solutions involving various user groups and centring development of their needs; and (4) technological development, which serves all other areas looking into using interoperable technological components. These areas are the main starting point for our conceptual framework.

To combine the four GovTech areas (Figure 6.2) with DT principles (Table 6.1) and practices (Table 6.2), we have analysed the ways in which DT is recognised to help overcome GovTech barriers in the literature. Given the scant literature discussing the specific connection between DT and GovTech, to build this understanding we have considered two main streams of literature. On one hand, we have analysed notions developed in the field of e-government and e-participation from 2000 to 2022 (Macintosh, 2004; Medaglia, 2007; Sæbø et al., 2008; Koussouris et al., 2011; Medaglia, 2012; Reddick and Norris, 2013; Wirtz et al., 2018; Santamaría-Philco et al., 2019; Adnan et al., 2022) performing a review of grey and scientific literature to capture the foundational, consolidated as well as emerging topics. Notions have been identified by reviewing theoretical frameworks and case studies, thus extracting the most relevant ways (principles and practices) in which DT can contribute. On the other hand, we have selected the contributions in the DT literature that have specifically linked design and public sector innovation. In accordance with qualitative rigour in inductive research (Eisenhardt, 2021), we clustered all the notions identified through this process. We then iterated the insights extracted from the literature in two main ways. First, we included considerations coming from our direct experience in numerous research projects working on this topic (starting from 2011 until today, we conducted over 20 EU funded projects). Then, we further tested and refined our framework through dialogue with other researchers (10 overall), expert in applying design to public sector innovation. The results of this analysis are presented and discussed in paragraph 5.

6.5 HOW DT PRINCIPLES AND PRACTICES CAN HELP REINFORCE GOVTECH

Our framework (Figure 6.3) presents and interrelates the relationship between DT principles and practices and GovTech focus areas, underscoring the direct interdependencies and influences that DT has on each element. In particular, it highlights how DT principles horizontally inform any activity performed when developing GovTech solutions. Furthermore, it adapts DT

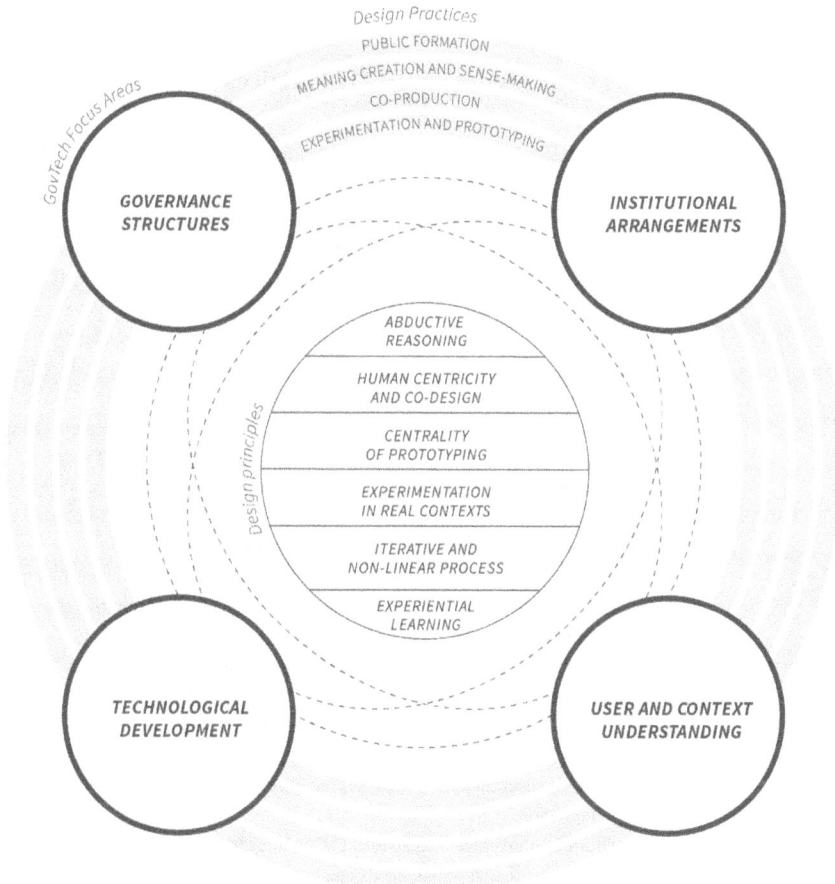

FIGURE 6.3 Conceptual framework for DT in GovTech: Interdependencies among DT principles, DT practices, and GovTech focus areas.

practices to align with and support each of the four GovTech focus areas in different ways.

As a complement to the framework, Table 6.4 further details how the DT practices identified in the literature contribute to each area, enhancing the alignment of the GovTech domain with a multi-level perspective that ranges from the inward-facing activities (organisational processes and governance structures) to the outward-facing ones (service development and delivery).

The following paragraphs elaborate further on our framework, providing a comprehensive approach that can help bridge theoretical concepts from two domains, with an operational perspective, delving into how DT can support GovTech in terms of governance structures, institutional arrangements, user and context understanding, and technological development.

TABLE 6.4

GovTech Focus Areas Combined with the DT Practices

GovTech Focus Area	Relevance for Each DT Practices Against the GovTech Focus Area
1 Governance structures	**Publics formation** enhances governance structures by fostering inclusivity and democratic participation in decision-making. By integrating multiple perspectives and promoting public debates, it helps shape more robust governance frameworks that are responsive to diverse community needs.
	Meaning creation and sense-making aid in interpreting complex governance challenges and policies. It can inform policymakers by providing deep insights into stakeholder perspectives, thus supporting more informed governance decisions.
	Co-production aligns with creating collaborative governance structures where citizens and government officials co-create policies and services. This enhances transparency and trust, leading to governance that is co-managed by its stakeholders.
	Experimentation and prototyping allows for testing new regulatory frameworks and governance models in a controlled, iterative manner. It helps refine governance structures to ensure they are effective before full-scale implementation.
2 Institutional arrangements	**Publics formation** encourages the engagement of diverse groups in the policymaking process, ensuring that institutional frameworks (e.g. laws, regulations, etc.) are representative and equitable.
	Meaning creation and sense-making helps institutions consider societal values and expectations as central inputs for policy making and implementation.
	Co-production leverages engagement of citizens in the policy development process to ensure that public needs and expectations are considered and relevant.
	Experimentation and prototyping allows institutions to trial and refine their arrangements in an experimental setting, engaging citizens for gathering insights on how to improve institutional policies and structures.
3 User and context understanding	**Publics formation** helps engage citizens and experts in public dialogue aimed at unpacking relevant and contrasting points of views on societal issues.
	Meaning creation and sense-making translate user data and behaviours into actionable insights that generate value to inform the design and implementation of public services.
	Co-production engages users directly in the creation of solutions, ensuring that their needs and contexts are fully understood and integrated into the service development process.
	Experimentation and prototyping with users in real-world contexts provides understanding on how they interact with solutions, allowing for iterative improvements based on actual usage and feedback.

(*Continued*)

TABLE 6.4 (*Continued*)
GovTech Focus Areas Combined with the DT Practices

GovTech Focus Area	Relevance for Each DT Practices Against the GovTech Focus Area
4 Technological development	**Publics formation**, while less directly connected, engages expert users and stakeholders in dialogues that influence technological development, leading to more desirable solutions.
	Meaning creation and sense-making help define clear, meaningful technology objectives that resonate with users and stakeholders, ensuring technology solutions are relevant and valuable.
	Co-production involves users and stakeholders in the design process, leading to technology that is more likely to meet real-world needs and be readily adopted.
	Experimentation and prototyping allows for iterative testing and refinement of technologies before they are fully deployed, reducing risks and enhancing the efficacy of technological solutions.

6.5.1 DT FOR GOVERNANCE STRUCTURES

The concept of governance design, as explained by Koppenjan and Groenewegen (2005), involves setting collective goals, assigning decision rights, and managing the coordination of GovTech's design, implementation, and operations. It transcends simple decision-making to encompass comprehensive change management processes. These processes aim to clarify participant roles within the GovTech lifecycle, from policy initiation to technology deployment and beyond. Common methodologies applied in reshaping governance include engaging in multi-helix collaboration (Leydesdorff, 2012), initiating public-private collaboration (Klievink et al., 2016), or establishing GovTech incubation programs (Kuziemski et al., 2022). Such methods often lead to defining governance models, new decision-making bodies or implementation roadmaps. However, these methods do not address well enough the need to co-develop GovTech governance structures and maximise the potential, yet minimise the risks of GovTech. Currently, we lack guidance in effectively steering the design and governance of GovTech solutions. This includes the installation of decision-making bodies, working groups and communication structures. Further to this, public agencies often lack the absorptive capacity (Cohen and Levinthal, 1990) to uptake new governance mechanisms and technology quickly: they need upskilling to examine GovTech solutions and procure effective co-creation, experimentation and implementation activities (OECD, 2017; 2024). Resource and capacity constraints can limit the extent to which public agencies can proactively shape GovTech solutions. Finally, the political fear of making mistakes, which can follow from experimentation, is deeply rooted in the public sector and can result in inaction or insufficient leadership commitment (Klievink et al., 2016).

The development of governance structures that are appropriate for GovTech can be supported by various DT practices such as co-production and publics formation, which can help ensure transparency and foster stakeholder trust. These structures support continuous interaction between citizens and government, crucial for refining governance models through practices like experimentation and prototyping. These iterative practices test and improve regulatory frameworks and governance models, aligning them more closely with public needs before they are fully implemented. An example can be seen in the development and deployment of digital platforms for urban mobility, such as those used in the Smart Dublin initiative (Cardullo and Kitchin, 2019). In this project, the Dublin City Council collaborated with technology companies, local universities, and citizens to create a comprehensive digital platform aimed at improving urban transportation. Through a series of hackathons, public consultations, and co-design workshops, stakeholders from various sectors contributed ideas and feedback on the platform's features and functionalities. This helped enhance democratic participation, allowing for a multiplicity of viewpoints to shape more adaptive and responsive solutions. Extending this reasoning, DT can support governance processes by steering engagement and co-design with all interested stakeholders, thus embracing a more dynamic, inclusive, and experimental approach. Engaging multiple actors in governance also means moving beyond traditional silos to collaborate across different sectors and disciplines, often also in piloting or real-life experimentation settings to test innovation in safer contexts where input and feedback nurture implementation before its broader application. Such practices require a cultural shift towards openness, flexibility, and a tolerance for failure and iterative learning that DT promises to bring. Together, DT practices can help ensure that governance structures are not only robust but also reflective of and responsive to the diverse needs of the community they serve.

6.5.2   DT FOR INSTITUTIONAL ARRANGEMENTS

The processes linked to establishing institutional arrangements (or laws, regulations, norms, policies, and organisational structures), that are appropriate for GovTech, require establishing trust frameworks and defining clear roles and responsibilities, which help mitigate risks and guide the behaviour of all actors involved. These arrangements are also pivotal for clarifying who gets involved in decision-making processes, setting the terms for participation, and detailing the commitments from all parties. As noted by Janssen et al. (2017), these arrangements span both relational and contractual governance mechanisms, essential for maintaining a level-playing field and ensuring equitable contributions from all stakeholders. Establishing such arrangements is fraught with challenges due to the 'institutional void' highlighted by Hajer (2003). The novelty of GovTech as a policy field means that existing institutional structures often lack the specificity and adaptability needed to support the rapid evolution of technology and public expectations. Therefore, a significant challenge lies in developing institutions that not only manage the current needs but are also agile enough to adapt to future demands and technologies. This involves rigorous institutional (gap) analysis, legal engineering, and policy standardisation activities, which aim to develop trust frameworks, set norms, and establish audit and certification programs that underpin GovTech initiatives. Furthermore, the interaction between public agencies and private tech firms introduces complexity, often characterised by a mutual scepticism that can hinder collaboration. Establishing trust anchors is essential for building confidence between these entities, involving clear legal, data, and cybersecurity standards that ensure all parties adhere to agreed principles and practices. Addressing these institutional challenges is not merely about creating suitable institutional arrangements but also about enabling public agencies to effectively engage with and manage GovTech solutions. This includes enhancing their absorptive capacity to understand, adapt, and implement new technologies effectively. The future of institutional design in GovTech thus rests on the ability to foster robust partnerships, enhance transparency, and build a governance model that supports continuous innovation and mutual trust (Mergel et al., 2022).

In this GovTech area, DT can play a role in helping orchestrate the complex interplay of various stakeholders through trust building and process definition. The foundational aspect of institutional design in GovTech concerns establishing coherent 'rules of the game' (Bharosa, 2022) – a blend of formal laws, informal norms, and both public and private regulatory frameworks – that govern interactions and ensure coordinated actions. Such frameworks structure the actions within the GovTech ecosystem, balancing the input from various actors to foster accountability and shared values. An example of this is the development of the 'Digital Identity Framework' in Estonia which was collaboratively developed to identify key needs and concerns, user-friendliness, and adaptability to future technological changes. Following this example, the practice of publics formation can aid the policymaking process in several ways: by ensuring diverse groups are represented, it promotes equity and fairness. In addition, through the creation of meaning and sense-making, institutions can become more aware of and integrate societal values and expectations, making policies more relevant and impactful. Co-production helps ensure that institutional arrangements reflect the broad spectrum of community perspectives and requirements, thus increasing their likelihood of acceptance and fostering sense of ownership. Finally, experimentation and prototyping allow institutions to test and refine these arrangements in controlled settings, gathering valuable feedback to enhance policy effectiveness and adaptability. By introducing solutions into real-world experimental settings – although limited and potentially not fully representative – DT can enable the identification of possible negative outcomes prior to 'going public'. Similarly, prototyping provides a tangible version of the solution, facilitating dialogue among diverse stakeholders. This hands-on approach not only sparks meaningful discussions but also garners essential insights triggered by direct interaction with the prototypes, enhancing the development process through collaborative feedback. In this discourse, the inclusion of multiple perspectives during co-producing, experimenting, and prototyping is a way to gain a more thorough coverage of potential impacts.

6.5.3 DT FOR USER AND CONTEXT UNDERSTANDING

In the GovTech context, a clear and comprehensive understanding of the specificities of users and the environments where a solution will operate can largely affect the design and delivery of meaningful solutions. This focus can ensure that GovTech solutions not only meet but exceed the expectations of public services. Key objectives include deeply understanding the diverse user groups and their unique needs, actively involving these users in the design and implementation processes of GovTech, and ensuring that the solutions developed are safe and uphold human values (Mariani et al., 2023) – aligning with the necessity to instill ethical values into technology (Luppicini, 2012; van de Poel, 2013). To achieve these goals, methods such as co-creation and citizen engagement are extensively adopted (Concilio et al., 2022), allowing for participatory approaches where inputs and feedback are solicited directly from users at various stages of the development process. Techniques like value-driven design and participatory risk assessment are also employed to ensure that technologies align with ethical standards and are inherently safe and desirable for users. Moreover, approaches like journey mapping are instrumental in identifying potential pitfalls and enhancing the user experience by assessing both positive and negative user scenarios, thereby preparing for and mitigating possible adverse outcomes. Addressing the challenges in this domain involves fostering a high level of democratic engagement, where various stakeholders including public agencies, private sector entities, and user communities collaborate. This not only amplifies the legitimacy of GovTech solutions but also ensures that a broad spectrum of perspectives and values are embedded right from the early stages of solution development and technology deployment.

Despite this being well understood in the theory, the field still grapples with determining the most effective ways to engage diverse user groups and integrate their inputs systematically. For instance, a significant challenge lies in designing responsibly within the GovTech space, particularly concerning privacy, security, fairness, and ethical considerations. Given the complex data environments and the involvement of multiple actors, ensuring that responsibilities are clearly defined and managed is critical. The 'problem of many hands' (Bovens, 1998) often complicates accountability and responsibility, especially when numerous components and actors contribute to a single GovTech solution. DT is again instrumental here. Considering the example of the UK's 'FixMyStreet' platform which enables citizens to report local issues, extensive user research has been used to understand needs and behaviours of residents before implementing the platform. Further, co-creation workshops have been held to co-design with all stakeholders while residents have directly contributed to implementing parts of the platform. Finally, several experimentations have been run before deployment (Baykurt, 2011). Such iterative design-led process has helped ensure that the final solution was not only effective in addressing local issues but also widely accepted and trusted by the community.

6.5.4 DT FOR TECHNOLOGICAL DEVELOPMENT

In the technological development domain of GovTech, the design and deployment of digital solutions necessitate a comprehensive array of components (technological and human) working in harmony (Kysh, 2022). These components vary from tightly integrated in-house applications to modular systems that use APIs for broader accessibility and functionality. The overarching objective in this area is to develop interoperable digital components that align with the human-centric, institutional and governance frameworks established within a broader GovTech strategy (Bharosa, 2022). Approaches that can be employed here include user experience design, agile development, and rapid prototyping, which facilitate the iterative development and testing of prototypes, with subsequent cycles of implementations, where prototypes can range from conceptual designs (i.e. system architectures and process flow diagrams) to more tangible outputs (i.e. apps, data ledgers, and APIs).

Despite these approaches, challenges persist, particularly in ensuring the interoperability of various technological components. For instance, specific service providers might manage digital identities, while public agencies handle essential citizen data, and other specialised market parties offer on-demand data services. The complexity of integrating these diverse elements requires a unified approach to data sharing, standardisation and compliance with open standards for interoperability useful for enabling seamless interactions between disparate systems and technologies (Dener et al., 2021). By harnessing the insights of expert users who understand the intricacies of these challenges, DT can collect their inputs for steering the development of functionalities that are both innovative and practical. Such inputs ensure that solutions are fine-tuned to improve accuracy and responsiveness, especially when handling personal and sensitive data. DT can be introduced to analyse data interaction patterns, hence informing the development of more intuitive data management systems that serve both internal government needs and public usability standards. For instance, co-developing a public-facing digital service with both IT developers, civil servants, and citizens can ensure that the service is accessible, user-friendly, and meets the diverse needs of a broad user base, including minorities and persons with special needs. This approach is exemplified by Finland's Immigration Service (Migri), which introduced a virtual assistant (chatbot) named 'Kamu' to reduce administrative burdens and improve the immigrant experience with public services (Swan, 2018; Komatsu et al., 2021). In order to guarantee that the chatbot's content was relevant and would actually reduce the number of phone calls, the bot was co-designed with the customer service staff who emphasised the need for the bot to be trustworthy and transparent. In addition, the interaction with migrants allowed the tailoring of the 'Kamu' chatbot, giving it a tone of voice and a way of interacting with migrants that was designed and programmed to be supportive and pleasant (Komatsu et al., 2021).

In this area the path-dependence of government agencies often also presents significant challenges related to transitioning from existing legacy systems. Expanding on this, the rigidity of existing infrastructures can hinder the incorporation of new solutions, compelling agencies to devise strategies that connect older systems with emerging technologies. This effort aims to ensure continuity and avoid disruptions in service while transitioning to more advanced technological platforms. Addressing quality concerns is another critical aspect, as data, system, and service quality must meet high standards to ensure the reliability and security of GovTech solutions. These quality dimensions include the accuracy, completeness, timeliness, and relevance of data; the integrity, availability, and security of systems; and the responsiveness and user-friendliness of services. In this sense, refining GovTech solutions through DT practices helps to work in a risk-managed environment allowing multiple stakeholders to see and interact with proposed solutions early in the development process and contribute to their refinement, enhancing relevance and novelty. For instance, pilot testing of solutions related to digital identity verification systems can take place in selected communities, engaging its participants to identify usability issues and security vulnerabilities before nationwide rollout.

Ultimately, the success of technological development in GovTech depends on the ability to manage these challenges effectively, ensuring that digital solutions not only meet current regulatory and functional requirements but are also adaptable to future changes and innovations. This involves a continuous process of testing, feedback, and refinement, which is essential for developing robust, efficient, and user-friendly GovTech solutions and also inherent in the DT approach explained in our framework.

6.6   A WAY FORWARD: LEARNING AND ORGANISATIONAL CHANGE

In the evolving landscape of GovTech, the discussion on integrating design culture into public sector innovation reveals a progressive shift from merely focusing on the structure of services to emphasising organisational transformation. Organisational change can be triggered by embedding design knowledge within public organisations to cultivate a culture that is open to exploration and innovation, centres on user needs and values experimentation and iteration (Deserti and Rizzo, 2014; Kimbell, 2015; Tõnurist et al., 2017; Blomkamp, 2018; Lewis et al., 2020). The integration of DT principles and practices discussed throughout this chapter ultimately aims at supporting this more profound shift by reinforcing the four GovTech focus areas and promoting a diffused human-centric approach to developing innovative digital public services. As discussed in the literature (Allio, 2014; Lewis et al., 2020; Brinkman et al., 2023), adopting DT principles and practices at the organisational level can be critical for change in the public sector. Several scholars (Deserti and Rizzo, 2014; Elsbach and Stigliani, 2018; Komatsu, 2020; Steen et al., 2011) acknowledge that adopting DT to innovate within the public sector can catalyse two fundamental shifts: (1) a shift in outward-facing practices that aligns public services with societal needs and people's capabilities, such as transforming communication between public organisations and citizens; and (2) a shift in inward-facing processes to transforms internal processes and establishes collaborations that can help public organisations deliver on their promises to society, for instance adopting co-production strategies and experimentation (Rizzo et al., 2017; Mariani et al., 2023).

These inward-facing change strategies also highlight the importance of balancing top-down policy directives with bottom-up employee practices and service implementation – a widely discussed challenge still seeking reconciliation in practice (Junginger, 2013; Bason, 2017; Clarke and Craft, 2019; Blomkamp, 2022). Working on both these dimensions can mark the beginning of meaningful organisational change that permeates all the processes and departments of a public institution through establishing cycles of action and reflection (Argyis and Schon, 1978). Concretely, this involves initially engaging in new and challenging tasks that push employees out of their comfort zones (*reflection in action*). Subsequent reflection on these actions (*reflection on action*) can then foster a renewal of practices and processes aimed at long-term change. Scholars across organisation studies, management, and design studies (Liedtka, 2011; Mazzucato, 2011; Borins, 2014; Bourgon, 2015; Magalhães and Proper, 2017), argue that such processes can be supported through cycles of experiential learning (Kolb, 1984a) where *reflection in action* focuses on revising organisational processes, while *reflection on action* can generate 'sticky knowledge' that pushes the organisation towards innovation. In other words, this alternation of phases encourages the organisation to experiment with new practices like DT, with the goal of challenging and revising established mindsets for the long term. In this context, design research and studies provide a transformative perspective on organisational evolution by highlighting the roles of design principles and practices in reshaping organisational structures. This emerging understanding of design is progressively guiding public organisations and governments as they rethink and revise their operational strategies (Junginger, 2017).

Our framework builds on this discourse, underscoring that embracing DT within the public sector, and particularly in GovTech, involves extending beyond mere user-centeredness. It necessitates the establishment of new democratic and participatory settings that challenge and disrupt the traditional dichotomy between public and private sectors as well as the conventional notion of public-private partnerships, advocating for the establishment of an ecosystem as a collaborative continuum that fosters transformative change across all levels of governance and service delivery. Here, technology serves merely as a means to an end: the suffix 'Tech' in the acronym GovTech is an excuse to move towards the long-needed transformation into entities that are more flexible, transparent, and innovation-oriented.

6.7 CONCLUSIONS

GovTech remains an emerging and evolving notion both in theory and practice, requiring deeper exploration to fully comprehend how its principles can be effectively operationalised. There is a distinct need for further research to delineate its meaning and features from the related concepts of eGovernment and Digital Government, and to explore how to nurture practical strategies for its development in practice. To bridge the existing gap between theoretical frameworks and practical application, this chapter has provided an examination of existing definitions and the recognition of how DT principles and practices can concretely and comprehensively support the reinforcement of the four main GovTech focus areas. In addition, the chapter has introduced a framework highlighting the relevance of adopting DT, discussing the need to go beyond confined experimentation and limited adoption and implementation. To avoid leaving societal needs behind in the development of GovTech, we have also discussed the critical requirement for wider organisational change, with DT acting as an innovative approach capable of aligning economic, societal, and cultural needs. By grounding innovation in human-centric principles and practices, the introduction of DT in this context seeks to not only to develop better solutions and steer innovation, but also to create public value that can renovate the relationship between government and society (Nose, 2023).

NOTES

1 https://commission.europa.eu/strategy-and-policy/priorities-2019-2024/europe-fit-digital-age/europes-digital-decade-digital-targets-2030_en
2 https://joinup.ec.europa.eu/collection/govtechconnect
3 The main principles for the successful implementation of digital public services are listed as: digital by default, citizen-centricity, inclusiveness, trustworthiness, accessibility, openness, transparency and interoperability.

REFERENCES

Adnan, Mohammed, Masitah Ghazali, and Nur Zuraifah Syazrah Othman. 2022. "E-Participation within the Context of e-Government Initiatives: A Comprehensive Systematic Review." *Telematics and Informatics Reports* 8 (December):100015. https://doi.org/10.1016/j.teler.2022.100015.

Allio, Lorenzo. 2014. *Design Thinking for Public Service Excellence*. UNDP Global Centre for Public Service Excellence, Singapore. https://www.undp.org/publications/designthinking-public-service-excellence.

Alshuwaikhat, Habib M, and Danjuma I Nkwenti. 2002. "Visualizing Decision-making: Perspectives on Collaborative and Participative Approach to Sustainable Urban Planning and Management." *Environment and Planning B: Planning and Design* 29 (4): 513–31. https://doi.org/10.1068/b12818.

Argyis, C, and D Schon. 1978. *Organizational Learning*. Addison-Wesley, Reading.

Argyris, Chris, and Donald A Schön. 1978. *Organizational Learning: A Theory of Action Perspective*. Vol. 1. Addison-Wesley, Reading.

Bason, Christian. 2017. *Leading Public Design: Discovering Human-Centered Governance*.

Baykurt, Burcu. 2011. "Redefining Citizenship and Civic Engagement: Political Values Embodied in FixMyStreet.Com." *AoIR Selected Papers of Internet Research* 1(0): 1–18. https://spir.aoir.org/ojs/index.php/spir/article/view/8811.

Beckman, Sara L., and Michael Barry. 2007. "Innovation as a Learning Process: Embedding Design Thinking." *California Management Review* 50 (1): 25–56. https://doi.org/10.2307/41166415.

Bekkers, VJJM, Lars G Tummers, and William H Voorberg. 2013. *From Public Innovation to Social Innovation in the Public Sector: A Literature Review of Relevant Drivers and Barriers.* Erasmus University, Rotterdam.

Bharosa, Nitesh. 2022. "The Rise of GovTech: Trojan Horse or Blessing in Disguise? A Research Agenda." *Government Information Quarterly* 39 (3): 101692. https://doi.org/10.1016/j.giq.2022.101692.

Bharosa, Nitesh, and Marijn Janssen. 2020. "Digicampus: Preliminary Lessons from a Quadruple Helix Ecosystem for Public Service Innovation." In *Ongoing Research, Practitioners, Posters, Workshops, and Projects of the International Conference EGOV-CeDEM-ePart*, 2797:195–203. https://ceur-ws.org/Vol-2797/paper19.pdf.

Björgvinsson, Erling, Pelle Ehn, and Per-Anders Hillgren. 2012. "Agonistic Participatory Design: Working with Marginalised Social Movements." *CoDesign* 8 (2–3): 127–44. https://doi.org/10.1080/15710882.2012.672577.

Blomkamp, Emma. 2018. "The Promise of Co-Design for Public Policy." *Australian Journal of Public Administration* 77 (4): 729–43. https://doi.org/10.1111/1467-8500.12310.

Blomkamp, Emma 2022. "Systemic Design Practice for Participatory Policymaking." *Policy Design and Practice* 5 (1): 12–31. https://doi.org/10.1080/25741292.2021.1887576.

Bogers, Marcel, Allan Afuah, and Bettina Bastian. 2010. "Users as Innovators: A Review, Critique, and Future Research Directions." *Journal of Management* 36 (4): 857–75. https://doi.org/10.1177/0149206309353944.

Borins, Sandford F. 2014. *The Persistence of Innovation in Government.* Brookings Institution Press. http://www.jstor.org/stable/10.7864/j.ctt6wpcpq.

Bostrom, Robert P., and J. Stephen Heinen. 1977. "MIS Problems and Failures: A Socio-Technical Perspective. Part I: The Causes." *MIS Quarterly* 1 (3): 17–32. https://doi.org/10.2307/248710.

Bourgon, Jocelyn. 2015. *Public Innovation and Public Purpose. A Follow Up to the OECD Conference, Innovating the Public Sector: From Ideas to Action.* Ottawa: Public Governance International. https://www.pgionline.com/wp-content/uploads/2015/08/Public-Innovation-and-Public-Purpose.pdf.

Bovens, Marcus Alphons Petrus (1998) *The Quest for Responsibility: Accountability and Citizenship in Complex Organisations.* Cambridge university press, Cambridge.

Brinkman, Geert, Arwin van Buuren, William Voorberg, and Mieke van der Bijl-Brouwer. 2023. "Making Way for Design Thinking in the Public Sector: A Taxonomy of Strategies." *Policy Design and Practice* 6 (3): 241–65. https://doi.org/10.1080/2574 1292.2023.2199958.

Brown, Tim (2009) *Change by Design.* HarperCollins, New York.

Cardullo, Paolo, and Rob Kitchin. 2019. "Smart Urbanism and Smart Citizenship: The Neoliberal Logic of 'Citizen-Focused' Smart Cities in Europe." *Environment and Planning C: Politics and Space* 37 (5): 813–30. https://doi.org/10.1177/02637 74X18806508.

Clarke, Amanda, and Jonathan Craft. 2019. "The Twin Faces of Public Sector Design." *Governance* 32 (1): 5–21. https://doi.org/10.1111/gove.12342.

Cohen, Wesley M, and Daniel A. Levinthal. 1990. "Absorptive Capacity. A New Perspective on Learning and Innovation." *Administrative Science Quarterly* 35 (1): 128–52. http://www.jstor.org/stable/2393553.

Concilio, Grazia, Giuliana Costa, Maryam Karimi, Maria Vitaller del Olmo, and Olga Kehagia. 2022. "Co-Designing with Migrants' Easier Access to Public Services: A Technological Perspective." *Social Sciences* 11 (2). https://doi.org/10.3390/socsci11020054.

Concilio, Grazia, Maryam Karimi, and Ilaria Mariani. 2022. "A Wiki-Space Driven Approach towards Inter-Organizational Learning." In *IFKAD – 17th International Forum on Knowledge Asset Dynamics*, 1748–61. Lugano, Switzerland.

Coughlan, Peter, Jane Fulton Suri, and Katherine Canales. 2007. "Prototypes as (Design) Tools for Behavioral and Organizational Change: A Design-Based Approach to Help Organizations Change Work Behaviors." *The Journal of Applied Behavioral Science* 43 (1): 122–34. https://doi.org/10.1177/0021886306297722.

Dantec, Le, and A Christopher (2016) *Designing Publics*. MIT Press, Cambridge.

Dener, Cem, Hubert Nii-Aponsah, Love E. Ghunney, and Kimberly D. Johns. 2021. *GovTech Maturity Index: The State of Public Sector Digital Transformation*. International Development in Focus. The World Bank. https://doi.org/10.1596/978-1-4648-1765-6.

Deserti, Alessandro, and Francesca Rizzo. 2014. "Design and Organizational Change in the Public Sector." *Design Management Journal* 9 (1): 85–97. https://doi.org/10.1111/dmj.12013.

Deutschmann, Maiia, and Moritz Botts. 2015. "Experiential Learning Through the Design Thinking Technique." In *The Palgrave Handbook of Experiential Learning in International Business*, edited by Vas Taras and Maria Alejandra Gonzalez-Perez, 449–63. Palgrave Macmillan UK, London. https://doi.org/10.1057/9781137467720_26.

Dijk, Niels van, Simone Casiraghi, and Serge Gutwirth. 2021. "The 'Ethification' of ICT Governance. Artificial Intelligence and Data Protection in the European Union." *Computer Law & Security Review* 43 (November):105597. https://doi.org/10.1016/j.clsr.2021.105597.

DiSalvo, Carl. 2010. "Design, Democracy and Agonistic Pluralism." In *DRS2010 RESEARCH PAPERS*. Montreal, Canada. https://dl.designresearchsociety.org/drs-conference-papers/drs2010/researchpapers/31.

Dong, Andy, and Erin MacDonald. 2017. "From Observations to Insights: The Hilly Road to Value Creation." In *Analysing Design Thinking: Studies of Cross-Cultural Co-Creation*, by Bo T. Christensen, Linden J. Ball, and Kim Halskov, 465–82. CRC Press, London.

Dorst, Kees. 2011. "The Core of 'Design Thinking' and Its Application." *Interpreting Design Thinking* 32 (6): 521–32. https://doi.org/10.1016/j.destud.2011.07.006.

Dorst, Kees, and Nigel Cross. 2001. "Creativity in the Design Process: Co-Evolution of Problem–Solution." *Design Studies* 22 (5): 425–37. https://doi.org/10.1016/S0142-694X(01)00009-6.

Drews, Christiane. 2009. "Unleashing the Full Potential of Design Thinking as a Business Method." *Design Management Review* 20 (3): 38–44. https://doi.org/10.1111/j.1948-7169.2009.00020.x.

Durose, Catherine, and Liz Richardson. 2015. *Designing Public Policy for Co-Production: Theory, Practice and Change*. Policy Press, Bristol. https://doi.org/10.2307/j.ctt1t896qg.

Eisenhardt, Kathleen M. 2021. "What Is the Eisenhardt Method, Really?" *Strategic Organization* 19 (1): 147–60. https://doi.org/10.1177/1476127020982866.

Elsbach, Kimberly D., and Ileana Stigliani. 2018. "Design Thinking and Organizational Culture: A Review and Framework for Future Research." *Journal of Management* 44 (6): 2274–2306. https://doi.org/10.1177/0149206317744252.

European Commission – Directorate-General for Research and Innovation. 2013. "Powering European Public Sector Innovation: Towards a New Architecture. Report of the Expert Group on Public Sector Innovation." Research and Innovation. European Commission – Directorate-General for Research and Innovation. https://doi.org/10.2777/51054.

Evans, Mark, and Nina Terrey (2016) "Co-Design with Citizens and Stakeholders." *Evidence-Based Policy Making in the Social Sciences: Methods That Matter*, 243.

Filer, Tanya. 2019. "Thinking about GovTech A Brief Guide for Policymakers." *Bennett Institute for Public Policy. University of Cambridge.* https://www.bennettinstitute. cam.ac.uk/wp-content/uploads/2020/12/Thinking_about_Govtech_Jan_2019_online-1.pdf.

Grönroos, Christian, and Päivi Voima. 2013. "Critical Service Logic: Making Sense of Value Creation and Co-Creation." *Journal of the Academy of Marketing Science* 41 (2): 133–50. https://doi.org/10.1007/s11747-012-0308-3.

Hajer, Maarten. 2003. "Policy without Polity? Policy Analysis and the Institutional Void." *Policy Sciences* 36 (2): 175–95. https://doi.org/10.1023/A:1024834510939.

Holloway, Matthew. 2009. "How Tangible Is Your Strategy? How Design Thinking Can Turn Your Strategy into Reality." Edited by Vijay Kumar. *Journal of Business Strategy* 30 (2/3): 50–56. https://doi.org/10.1108/02756660910942463.

Janssen, Marijn, Haiko van der Voort, and Agung Wahyudi. 2017. "Factors Influencing Big Data Decision-Making Quality." *Journal of Business Research* 70 (January):338–45. https://doi.org/10.1016/j.jbusres.2016.08.007.

Junginger, Sabine. 2013. "Design and Innovation in the Public Sector: Matters of Design in Policy-Making and Policy Implementation." *Annual Review of Policy Design* 1(1): 1–11.

Junginger, Sabine. 2014. "Towards Policy-Making as Designing: Policy-Making Beyond Problem-Solving and Decision-Making." In *Design for Policy*, edited by Christian Bason, 57–69. Routledge, London. https://doi.org/10.4324/9781315576640.

Junginger, Sabine. 2017. "Design Research and Practice for the Public Good: A Reflection." *She Ji: The Journal of Design, Economics, and Innovation* 3 (4): 290–302. https://doi.org/10.1016/j.sheji.2018.02.005.

Khan, Anupriya, and Satish Krishnan. 2021. "Citizen Engagement in Co-Creation of e-Government Services: A Process Theory View from a Meta-Synthesis Approach." *Internet Research* 31 (4): 1318–75. https://doi.org/10.1108/INTR-03-2020-0116.

Kimbell, Lucy. 2011. "Rethinking Design Thinking: Part I." *Design and Culture* 3 (3): 285–306. https://doi.org/10.2752/175470811X13071166525216.

Kimbell, Lucy. 2012. "Rethinking Design Thinking: Part II." *Design and Culture* 4 (2): 129–48. https://doi.org/10.2752/175470812X13281948975413.

Kimbell, Lucy. 2015. *Applying Design Approaches to Policy Making: Discovering Policy Lab*. University of Brighton, Brighton.

Klievink, Bram, Nitesh Bharosa, and Yao-Hua Tan. 2016. "The Collaborative Realization of Public Values and Business Goals: Governance and Infrastructure of Public–Private Information Platforms." *Government Information Quarterly* 33 (1): 67–79. https://doi.org/10.1016/j.giq.2015.12.002.

Kolb, David A. 1984a. *Experience as the Source of Learning and Development*. Prentice Hall, Upper Sadle River.

Kolb, David A. 1984b. *Experiential Learning as the Science of Learning and Development*. Englewood Cliffs NPH.

Komatsu, Tamami Tiffany. 2020. "Transforming Public Sector Organizations through Design Culture: The Relationship between Design Practice, Innovation and Organizational Change." PhD Dissertation, Milano, Italy: Politecnico di Milano – Department of Design. http://hdl.handle.net/10589/164275.

Komatsu, Tamami Tiffany, Mariana Salgado, Alessandro Deserti, and Francesca Rizzo. 2021. "Policy Labs Challenges in the Public Sector: The Value of Design for More Responsive Organizations." *Policy Design and Practice* 4 (2): 271–91. https://doi.org/10.1080/25741292.2021.1917173.

Koppenjan, Joop, and John Groenewegen. 2005. "Institutional Design for Complex Technological Systems." *International Journal of Technology, Policy and Management* 5 (3): 240–57. https://doi.org/10.1504/IJTPM.2005.008406.

Koussouris, Sotirios, Yannis Charalabidis, and Dimitrios Askounis. 2011. "A Review of the European Union eParticipation Action Pilot Projects." Edited by Alexander Prosser. *Transforming Government: People, Process and Policy* 5 (1): 8–19. https://doi.org/10.1108/17506161111114617.

Kuziemski, Maciej, Peter Ulrich, Ines Mergel, and Amanda Martinez. 2022. "GovTech Practices in the EU." JRC Research Reports JRC128247. Luxembourg: Publications Office of the European Union. https://doi.org/10.2760/700544.

Kysh, Liudmyla. 2022. "Peculiarities of Govtech Technologies Implementation in the Public Administration System." *Scientific Journal of Polonia University* 53 (4). https://doi.org/10.23856/5321.

Lewis, Jenny M, Michael McGann, and Emma Blomkamp. 2020. "When Design Meets Power: Design Thinking, Public Sector Innovation and the Politics of Policymaking." *Policy & Politics* 48 (1): 111–30. https://doi.org/10.1332/0305573 19X15579230420081.

Leydesdorff, Loet. 2012. "The Triple Helix, Quadruple Helix, …, and an N-Tuple of Helices: Explanatory Models for Analyzing the Knowledge-Based Economy?" *Journal of the Knowledge Economy* 3 (1): 25–35. https://doi.org/10.1007/s13132-011-0049-4.

Liedtka, Jeanne. 2011. "Learning to Use Design Thinking Tools for Successful Innovation." *Strategy & Leadership* 39 (5): 13–19. https://doi.org/10.1108/10878571111161480.

Liedtka, Jeanne. 2015. "Perspective: Linking Design Thinking with Innovation Outcomes through Cognitive Bias Reduction." *Journal of Product Innovation Management* 32 (6): 925–38. https://doi.org/10.1111/jpim.12163.

Liedtka, Jeanne, and Tim Ogilvie. 2011. *Designing for Growth: A Design Thinking Tool Kit for Managers.* Columbia University Press, New York.

Liu, Helen K. 2021. "Crowdsourcing: Citizens as Coproducers of Public Services." *Policy & Internet* 13 (2): 315–31. https://doi.org/10.1002/poi3.249.

Luppicini, Rocci (2012) *Ethical Impact of Technological Advancements and Applications in Society.* IGI Global, Hershey.

Macintosh, Ann. 2004. "Characterizing E-Participation in Policy-Making." In *37th Annual Hawaii International Conference on System Sciences, 2004. Proceedings of The*, 10 pp. https://doi.org/10.1109/HICSS.2004.1265300.

Magalhães, Rodrigo, and Henderik A. Proper. 2017. "Model-Enabled Design and Engineering of Organisations and Their Enterprises." *Organizational Design and Enterprise Engineering* 1 (1): 1–12. https://doi.org/10.1007/s41251-016-0005-9.

Mariani, Ilaria, Maryam Karimi, Grazia Concilio, Giuseppe Rizzo, and Alberto Benincasa. 2023. "Improving Public Services Accessibility Through Natural Language Processing: Challenges, Opportunities and Obstacles." In *Intelligent Systems and Applications*, edited by Kohei Arai, 272–89. Springer International, Cham, Switzerland. https://doi.org/10.1007/978-3-031-16075-2_18.

Mariani, Ilaria, Marzia Mortati, and Francesca Rizzo. 2023. "Strengthening E-Participation through Design Thinking. Relevance for Better Digital Public Services." In *Proceedings of the 24th Annual International Conference on Digital Government Research*, 224–32. DGO'23. New York, NY, USA: Association for Computing Machinery. https://doi.org/10.1145/3598469.3598494.

Mariani, Ilaria, Francesca Rizzo, and Grazia Concilio. 2023. "Towards Better Public Sector Innovation. Co-Designing Solutions to Improve Inclusion and Integration." *Disegno Industriale Industrial Design*, no. DSI 1 (November), 10. https://doi.org/10.30682/diiddsi23t2j.

Mazzucato, Mariana. 2011. "The Entrepreneurial State." *Soundings* 49 (49): 131–42. https://doi.org/doi:10.3898/136266211798411183.

McGann, Michael, Emma Blomkamp, and Jenny M. Lewis. 2018. "The Rise of Public Sector Innovation Labs: Experiments in Design Thinking for Policy." *Policy Sciences* 51 (3): 249–67. https://doi.org/10.1007/s11077-018-9315-7.

Medaglia, Rony. 2007. "The Challenged Identity of a Field: The State of the Art of eParticipation Research." *Information Polity* 12 (3): 169–81. https://doi.org/10.3233/IP-2007-0114.

Medaglia, Rony. 2012. "eParticipation Research: Moving Characterization Forward (2006–2011)." *Government Information Quarterly* 29 (3): 346–60. https://doi.org/10.1016/j.giq.2012.02.010.

Mergel, Ines. 2016. "Agile Innovation Management in Government: A Research Agenda." *Open and Smart Governments: Strategies, Tools, and Experiences* 33 (3): 516–23. https://doi.org/10.1016/j.giq.2016.07.004.

Mergel, Ines, Noella Edelmann, and Nathalie Haug. 2019. "Defining Digital Transformation: Results from Expert Interviews." *Government Information Quarterly* 36 (4): 101385. https://doi.org/10.1016/j.giq.2019.06.002.

Mergel, Ines, Peter Ulrich, Maciej Kuziemski, and Amanda Martinez. 2022. "*Scoping GovTech Dynamics in the EU.*" JRC128093. JRC Research Reports. Luxembourg: Joint Research Centre. https://publications.jrc.ec.europa.eu/repository/handle/JRC128093.

Mintrom, Michael, and Madeline Thomas. 2018. "Improving Commissioning Through Design Thinking." *Policy Design and Practice* 1 (4): 310–22. https://doi.org/10.1080/25741292.2018.1551756.

Mortati, Marzia. 2022. "New Design Knowledge and the Fifth Order of Design." *Design Issues* 38 (4): 21–34. https://doi.org/10.1162/desi_a_00695.

Mortati, Marzia, Sabrina Bresciani, Eui Young Kim, and Sabine Junginger. 2023. "Changing Organizations and Policies: Equipping Design for Systemic Transformation." In *IASDR 2023: Life-Changing Design*, edited by Daniela De Sainz Molestina, Laura Galluzzo, Francesca Rizzo, and Davide Spallazzo, 1–15. Milano, Italy: DRS. https://doi.org/10.21606/iasdr.2023.890.

Mortati, Marzia, Ilaria Mariani, and Francesca Rizzo. 2023. "How Design Thinking Can Support the Establishment of an EU GovTech Ecosystem." In *IASDR 2023: Life-Changing Design*, edited by Daniela De Sainz Molestina, Laura Galluzzo, Francesca Rizzo, and Davide Spallazzo, 1–29. Milano, Italy: DRS. https://doi.org/10.21606/iasdr.2023.356.

Mortati, Marzia, Louise Mullagh, and Scott Schmidt. 2022. "Design-Led Policy and Governance in Practice: A Global Perspective." *Policy Design and Practice* 5 (4): 399–409. https://doi.org/10.1080/25741292.2022.2152592.

Mumford, Enid. 2006. "The Story of Socio-Technical Design: Reflections on Its Successes, Failures and Potential." *Information Systems Journal* 16 (4): 317–42. https://doi.org/10.1111/j.1365-2575.2006.00221.x.

Nose, Manabu. 2023. "Inclusive GovTech: Enhancing Efficiency and Equity Through Public Service Digitalization." *IMF Working Paper*, no. 2023/226.

OECD. 2017. "Government at a Glance 2017." Paris, France: OECD Publishing. https://doi.org/10.1787/gov_glance-2017-en.

OECD. 2024. "2023 OECD Digital Government Index." https://doi.org/10.1787/1a89ed5e-en.

Osborne, Stephen P, Zoe Radnor, and Kirsty Strokosch. 2016. "Co-Production and the Co-Creation of Value in Public Services: A Suitable Case for Treatment?" *Public Management Review* 18 (5): 639–53. https://doi.org/10.1080/14719037.2015.1111927.

Poel, Ibo van de. 2013. "Translating Values into Design Requirements." In *Philosophy and Engineering: Reflections on Practice, Principles and Process*, edited by Diane P Michelfelder, Natasha McCarthy, and David E. Goldberg, 253–66. Dordrecht: Springer Netherlands. https://doi.org/10.1007/978-94-007-7762-0_20.

Public. 2021. "The State of European GovTech. Key Themes and Players in the European GovTech Ecosystem." https://www.public.io/report-post/the-state-of-european-govtech.

Reddick, Christopher G, and Donald F Norris. 2013. "E-Participation in Local Governments: An Empirical Examination of Impacts." In *ACM International Conference Proceeding Series*, 198–204. https://doi.org/10.1145/2479724.2479753.

Rizzo, Francesca, Alessandro Deserti, and Onur Cobanli. 2017. "Introducing Design Thinking in Social Innovation and in the Public Sector: A Design Based Learning Framework." *European Public & Social Innovation Review* 2 (1): 127–43. https://doi.org/10.31637/epsir.17-1.9.

Rizzo, Francesca, Alessandro Deserti, Stefano Crabu, Melanie Smallman, Julie Hjort, Stephanie Joy Hansen, and Massimo Menichinelli. 2018. "Co-Creation in RRI Practices and STI Policies." D1.2. SISCODE Deliverables. Cordis. https://ec.europa.eu/research/participants/documents/downloadPublic?documentIds=080166e5bedc3a0d&appId=PPGMS.

Sæbø, Øystein, Jeremy Rose, and Leif Skiftenes Flak. 2008. "The Shape of eParticipation: Characterizing an Emerging Research Area." *Government Information Quarterly* 25 (3): 400–428. https://doi.org/10.1016/j.giq.2007.04.007.

Sanders, Elizabeth B-N, and Pieter Jan Stappers. 2014. "Probes, Toolkits and Prototypes: Three Approaches to Making in Codesigning." *CoDesign* 10 (1): 5–14. https://doi.org/10.1080/15710882.2014.888183.

Sangiorgi, Daniela. 2015. "Designing for Public Sector Innovation in the UK: Design Strategies for Paradigm Shifts." Edited by Ms. Deborah Cox, Dr. Lawrence Green, and Dr. Krzysztof Borodako. *Foresight* 17 (4): 332–48. https://doi.org/10.1108/FS-08-2013-0041.

Santamaría-Philco, Alex, José H. Canós Cerdá, and M. Carmen Penadés Gramaje. 2019. "Advances in E-Participation: A Perspective of Last Years." *IEEE Access* 7:155894–916. https://doi.org/10.1109/ACCESS.2019.2948810.

Seravalli, Anna, Mette Agger Eriksen, and Per-Anders Hillgren. 2017. "Co-Design in Co-Production Processes: Jointly Articulating and Appropriating Infrastructuring and Commoning with Civil Servants." *CoDesign* 13 (3): 187–201. https://doi.org/10.1080/15710882.2017.1355004.

Steen, Marc, Menno Manschot, and Nicole De Koning. 2011. "Benefits of Co-Design in Service Design Projects." *International Journal of Design, Vol 5, No 2 (2011)*. https://www.ijdesign.org/index.php/IJDesign/article/view/890/346.

Swan, Kristin. 2018. "Leading with Legitimacy in Government Design Labs." Master's Thesis, Tampere, Finland: Aalto University. https://urn.fi/URN:NBN:fi:aalto-201809105091.

Szkuta, Katarzyna, Roberto Pizzicannella, and David Osimo. 2014. "Collaborative Approaches to Public Sector Innovation: A Scoping Study." *Special Issue on: Selected Papers from the 10th Conference in Telecommunications, Media and Internet Techno-Economics* 38 (5): 558–67. https://doi.org/10.1016/j.telpol.2014.04.002.

Tallinn Declaration. 2017. "Tallinn Declaration on eGovernment at the Ministerial Meeting during Estonian Presidency of the Council of the EU on 6 October 2017." https://digital-strategy.ec.europa.eu/en/news/ministerial-declaration-egovernment-tallinn-declaration.

Tõnurist, Piret, Rainer Kattel, and Veiko Lember. 2017. "Innovation Labs in the Public Sector: What They Are and What They Do?" *Public Management Review* 19 (10): 1455–79. https://doi.org/10.1080/14719037.2017.1287939.

Trischler, Jakob, Timo Dietrich, and Sharyn Rundle-Thiele. 2019. "Co-Design: From Expert- to User-Driven Ideas in Public Service Design." *Public Management Review* 21 (11): 1595–1619. https://doi.org/10.1080/14719037.2019.1619810.

Ubaldi, Barbara, Enzo Maria Le Fevre, Elisa Petrucci, Pietro Marchionni, Claudio Biancalana, Nanni Hiltunen, Daniela Maria Intravaia, and Chan Yang. 2019. "State of the Art in the Use of Emerging Technologies in the Public Sector," no. 31. https://doi.org/10.1787/932780bc-en.

Venturini, Tommaso, Donato Ricci, Michele Mauri, Lucy Kimbell, and Axel Meunier. 2015. "Designing Controversies and Their Publics." *Design Issues* 31 (3): 74–87. https://doi.org/10.1162/DESI_a_00340.

Wirtz, Bernd W., Peter Daiser, and Boris Binkowska. 2018. "E-Participation: A Strategic Framework." *International Journal of Public Administration* 41 (1): 1–12. https://doi.org/10.1080/01900692.2016.1242620.

World Bank Group. 2020. "GovTech: The New Frontier in Digital Government Transformation." Available at: https://thedocs.worldbank.org/en/doc/805211612215188198-0090022021/original/GovTechGuidanceNote1TheFrontier.pdf

Yuan, Qianli, and Mila Gascó-Hernández (2019) "Open Innovation in the Public Sector: Creating Public Value Through Civic Hackathons." *Public Management Review* 23, 523–44.

Zapata, Enrique. 2020. "The GovTech Index 2020. Unlocking the Potential of GovTech Ecosystems in Latin America, Spain and Portugal." CAF – Development Bank of Latin America & Oxford Insights. http://scioteca.caf.com/handle/123456789/1580.

Zimmerman, John, Jodi Forlizzi, and Shelley Evenson. 2007. "Research through Design as a Method for Interaction Design Research in HCI." In *Proceedings of the SIGCHI Conference on Human Factors in Computing Systems*, 493–502. CHI '07. New York, NY, USA: ACM. https://doi.org/10.1145/1240624.1240704.

7 Design Thinking and Artificial Intelligence
A Framework for Analysis

Nick Kelly and Kazjon Grace

7.1 INTRODUCTION

Artificial intelligence (AI) systems have been used by designers for decades (Coyne et al., 1990). This usage has increased rapidly in recent years alongside the capabilities of AI. The current generation of models, which includes large language models (LLMs) and diffusion models combined across multiple modalities (including OpenAI's ChatGPT and Dall-E or Google's Gemini) already aid designers and design teams in synthesising information, producing visualisations, generating ideas, and much more (Sreenivasan & Suresh, 2024). AI, in the form of LLMs, is being used across all design disciplines including architecture (Pena et al., 2021; Bölek et al., 2023), interaction design (Schmidt et al., 2024), and engineering design (Yüksel et al., 2023) for diverse purposes. Alongside empirical research exploring the uses of AI within design thinking, there is a need to advance the *theory* about AI in design to guide such work (Cash, 2018) and to conceptualise how humans and AIs work together in design thinking. Given the use of AI in design, how are we to conceptualise what has been referred to as *hybrid design thinking* (Wendrich, 2013) and *computational co-creativity* (Karimi et al., 2018), in which cognitive tasks are becoming increasingly distributed between human and AI agents?

In design, there has long been a relationship between design practitioners – those engaged in designerly way of knowing, doing, and acting in any domain – and the technology that supports their practice, whether that technology be a pencil or a computer. AI is changing this relationship: tools, as they have been conceived until now, have not possessed the level of design agency of generative AI (Lawton et al., 2023). A pencil influences how you draw, but it does not offer critique or suggestions. There is something that feels intuitively different about the use of AI in design compared to other technology used by designers; a qualitative difference rather than an incremental improvement of the technology that is available. This chapter aims to clarify why AI gives this feeling and provide a framework for analysing different configurations of humans and AIs involved in design activity. The chapter is structured by: (1) drawing on the literature to outline a model of design thinking which is grounded in situated cognition and is activity-centred that we refer to as a *frame-activity model of design thinking*; then

DOI: 10.1201/9781003487524-7

(2) outlining a framework for AI within design thinking based on this model; and finally (3) discussing the utility of this framework.

This framework is valuable for understanding and investigating the historic change that AI brings to those engaged in design thinking. As a tool for use within the design process, AI represents an incremental improvement on designers' workflows. As part of a design team with responsibility for framing design problems and designating activities, AI represents a break from all past uses of technology by designers. It is useful for those engaged in empirical studies of AI use in design as well as for those theorising how AI ought to be used in design.

7.2 A FRAME-ACTIVITY MODEL OF DESIGN THINKING

The term *design thinking* has its origins in decades of research into the cognitive activities of expert designers in their 'designerly thinking' (Cross, 2023). The intention in this section is to describe a model of design thinking that focuses on the cognitive activity of designers that allows them to commence a problem with a limited understanding of what they're doing (Dorst, 2011; Kelly & Gero, 2022). It does not aim to review or critique the concept of design thinking or widely used frameworks for design thinking (; Kimbell, 2011; Johansson-Sköldberg et al., 2013). Rather, it sets out the key elements of design thinking as the *frame* which defines the problem and (implicitly) limits the ideas that a designer has access to; and the *design activities* that a designer or design team undertakes to shift that frame in such a way that a satisfactory design solution becomes apparent.

7.2.1 CONCEPTS AND DESIGN COGNITION

Concepts and the relationships between them are the materials of all abstract thought (Murphy, 2004) and therefore the materials of design cognition. A design problem is not 'out there' waiting to be solved; it is definitional to the nature of design that designers must interpret a problem as a part of addressing it (Sosa et al., 2017). This interpretation requires the manipulation of concepts and their relationships. Similarly, a design solution must be conceptually formed in some way before it can be externalised as a design document.

Interpreting a design problem or conducting design activities requires a designer – or an AI system – to bring together concepts that represent that problem (whether or not they are aware that this is what they are doing). This is the creation of a *conceptual assemblage*, a cognitive structure composed of activated concepts and the relationships between them that pertain to this problem. This involves adopting a situated notion of design and of concepts, where those concepts may be embodied (Clancey, 1997; Barsalou, 2005).[1]

Of course, there are many such conceptual assemblages in cognitive activity, but not all of them constitute *design* cognition. Design activity involves a particular kind of conceptual assemblage, one that is oriented towards creating something that is currently unknown (the 'what' of design) that will achieve a desired goal (something of value); and where there is some way of working towards this 'what' given the many unknowns (the 'how') (Dorst, 2011). This kind of conceptual assemblage in design activity is known as the *design frame* and it is the basis for design abduction and other design activity (Dorst, 2015).

A design frame can be thought of as a conceptual space that includes both problem and solution spaces in designing (Maher & Poon, 1996; Dorst & Cross, 2001).[2] For example, this space might include an understanding of the stakeholders in a problem and their social relationships; an understanding of the physical spaces related to the problem and relationships between them; an understanding of relevant tools, technologies, and materials and relationships between them; a set of values; an understanding of the broader social, cultural, historical, political, and scientific contexts for the problem; relevant design histories; and so on. Kelly and Gero (2022) review the concept of design frames to suggest that:

> Designers, when addressing a design problem, conduct some cognitive activity that leads them to develop a set of concepts (where... that term includes values, beliefs, propositions, objects, symbols, etc.). A design frame is the singular name given to that conceptual structure, which is revealed through design moves/actions or utterances/representations.

(Kelly & Gero, 2022, p. 5)

7.2.2 Design Frames and Their Development

The design frame thus refers to the internal, conceptual world that arises in the context of the design problem; if there are multiple designers in a team then each has their own frame for the design activity. The word 'frame' here evokes the sense of being both a *scaffold* upon which something can be built (e.g., the frame for a house) as well as being a *border* within which something exists and that creates a limit (e.g., a picture frame) (Schön & Rein, 1994). A designer cannot reason about things that are outside the frame until the frame changes (e.g., in the example of the famous nine-dot problem[3]); nor can they commence design activity until they have some way of working, something to build upon (e.g., "let's approach this problem by first understanding its history and then talking to stakeholders").

One way to get a sense of the design frame that the designer is working with is to get them to externalise their frame. This might be done by asking them: "What is your current understanding of the problem that you're addressing and how are you trying to tackle it?" There are no effective research methods for directly studying the frame that a designer is working with: concepts within humans simply cannot be studied with available technology (though perhaps concepts within an AI could be, implementation dependent). Some methods have been developed for revealing aspects of framing by analysing design protocols (e.g., Valkenburg & Dorst, 1998 and subsequent studies using derivative methods), but these do not permit direct analysis of the specific concepts being used by a designer and their relationships.

It is clear from many protocol studies of designers that the understanding of a design problem – and the conception of the world of that problem – change during design activity (e.g., Suwa et al., 2000). This is analogous to saying that, during design activity, the design frame can change in many ways. Concepts can be added to or removed from the frame, or new connections between concepts made. For example, the designer may conduct research within the scope of a problem suggesting new ideas; or new ideas may emerge through a process of sketching and reflection. Concepts themselves can also change based on new knowledge. Attention can be placed on different parts of the design frame, changing the role played by other concepts and shifting the overall understanding. Concepts can be lost from the frame based on attention or memory.

One mark of an expert designer is knowing which design activities to undertake in order to change the frame in a desired way – developing this capacity is, arguably, the ultimate goal of all design education. The term design framing is used elsewhere to refer to the deployment of cognitive and metacognitive strategies that are "associated with useful development/refinement of a design frame" (Kelly & Gero, 2022, p. 14). We will refer to these actions that a designer takes to change the frame in a desired way as *design activities*. These activities emerge from within the design frame as a part of cognition.

7.2.3 DESIGN ACTIVITIES SHIFT DESIGN FRAMES

Design progresses through design activities and reflection in what has been theorised as the seeing-moving-seeing of design (Schon, 1983) or as cycles of learning and creating (Berkun, 2020). Design activities can operate on macro, meso, or micro scales and be formal or informal. For example, a formal design method like 'brainstorming as a design team' is a lengthy design activity, that may be run as a three-hour workshop with sticky notes and a range of tasks with a nominated facilitator. Less formal, off-the-cuff activities like 'sketching an idea', 'looking something up on Google', or 'talking to a stakeholder' also constitute design activities. This notion of an activity-centred analysis of designers is derived from the broad discourse of activity theory as it has been applied to design (Nardi, 1996; Zahedi et al., 2017).[4] Here we are concerned with bringing an account of design framing together with an activity-centred notion of design through a situated understanding of cognition.

Within a frame, a designer undertakes certain design activities, often without a clear idea of what their outcome will be or exactly why they are doing it. *In this model, design activities are analysed based on the way that they shift the design frame towards a more useful conceptual assemblage within which a desired design response can be found.* Where there are multiple designers in a team, each designer has their own frame, and it is through design activities that they influence each other (Dong et al., 2013). For example, each designer might externalise ideas through sketching, look at one another's sketches, and have a discussion about them. Through this process each designer's frame can shift; commonly, although not necessarily, they will become more aligned.

Any design activity can be analysed through the configuration of resources that they draw upon, with social, epistemic, and material being useful categories for understanding those resources (Goodyear & Carvalho, 2014). A design conversation between a designer and a colleague can be understood as an activity involving a social resource (with each adopting a role and using certain language). The tools that are used by designers in activities are typically material resources, such as model making with a knife, glue, ruler, and cardboard.

Activities may involve a configuration of resources where categorisation is unclear. Many computational tools, especially AI, involve the use of resources that can be considered through social, epistemic, and material lenses: the generation of a photo-realistic render of an idea using Dall-E is accessed through a cloud infrastructure and a computer (material); it occurs through a natural language dialogue between a designer and the AI (arguably social); and is motivated by an epistemic intention.

This model of design thinking as activities within a frame is suggested as an appropriate level of abstraction for theorising design thinking. It is consistent with extant models while sitting at a higher level of abstraction than most. It avoids the trap of modelling the design process based on typical grouping of activities that often occur together (e.g., "empathise, define, ideate, prototype, test" within the d.school model of design thinking), which trades predictive power for pedagogical utility. Instead, it highlights the notion that the role of activities should be understood in relation to how they change the design frame for individual designers, even when they are working within a team.

7.3   A FRAMEWORK FOR AI IN DESIGN THINKING

A framework for the use of AI in design thinking (Figure 7.1) can be articulated based upon the frame-activity model of design thinking. Within this framework, analysis of design thinking involves two questions: First, who gets to frame design activities? Second, what are the resources drawn upon in those activities? The activities themselves are depicted as a black box in Figure 7.1, as the

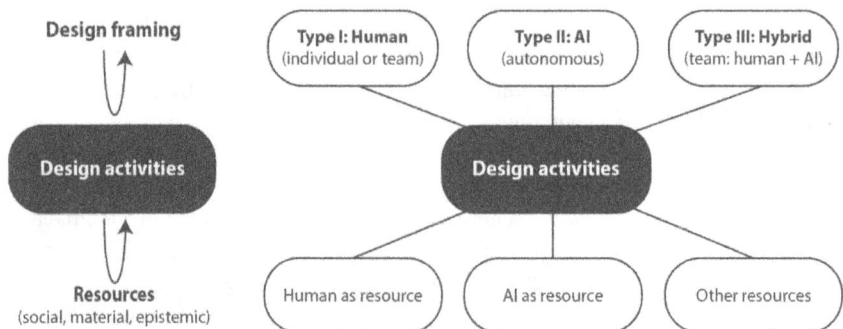

FIGURE 7.1   A framework for AI in design thinking.

content of activities can be anything as required by the frame. Arrows are loops to make it clear that activities occur through, and are dictated by, the design frame. Similarly, resources occur through, and are dictated by, activities.

The designer who is doing the framing is the one who gets to answer to the question "what is your current understanding of the design problem that you're addressing and the way that you're approaching it?" In a design team, this can be distributed amongst team members (be they human or AI) where team members influence each other's framing, devise activities together, and collaboratively evolve a collective design frame. In other design situations, stakeholders in the design process can be treated as mere resources within activities without having a role in the design framing. This is a common concern within the discourse on co-design, of how to ensure legitimate participation (Zelenko et al., 2021; Kerr et al., 2023). The framework will be elucidated by describing three 'types' of design thinking that involve humans and AIs working together. All three types can be described as 'hybrid' in the sense that humans and AIs are both involved, yet only type III is entirely hybrid in the sense that the design framing is shared.

7.3.1 Type I: Human Framing with AI Use in Activities

The first type of hybrid design thinking is between humans and AIs, with the human (or human team) acting as definer of the overall problem being solved and the designator of activities. AI is involved in designing in the context only of activities that have already been determined by the human(s).

The most common examples of hybrid design thinking today are type I. Examples involve AI carrying out research tasks for designers (e.g., reading books, surfing the web, and synthesising or summarising information, etc.) or carrying out generation tasks for designers (e.g., creating visualisations based upon prompts, iterating over possible ideas by suggesting multiple possibilities, interpolating data points, pretending to be a human for design testing, etc.). AI in this way is being used as an epistemic/material resource that draws upon epistemic resources. In the coming years, AI is likely to expand its ability to draw upon social and material resources within the context of a design activity.

A common example of Type I hybrid design thinking is to generate and refine ideas based upon a sequence of prompts. For example, Ploennigs and Berger (2023) detail the use of different public LLMs in the domain of architecture to produce "a workflow for creating images for interior designs and … views for exterior design that combines the strengths of the individual [LLMs]" (p. 1). The human is doing the design framing and drawing on the AI as a resource, hence it is type 1 (Figure 7.2).

For the purposes of disambiguation, it should be recognised that using presently available technology of OpenAI's GPT-5, a human designer can do all of the design activities for certain kinds of design problem through a few carefully selected prompts, including research, iterating, and refining ideas in multiple modalities, and generating some kind of design documents. To an external observer it may look as though the AI here has done a great deal of the 'design

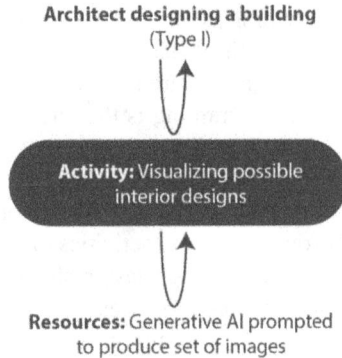

FIGURE 7.2   An architect designing draws upon AI as a resource as an example of type I hybrid design thinking.

work', yet the example still sits firmly within the territory of type I design thinking, in which the human possesses the key conceptual structures that define the problem, designate activities, and decide success criteria.

7.3.2   Type II: AI Framing with Human Use in Activities

AIs can theoretically do design without humans, something that has been theorised as *autonomous design* (Seidel et al., 2020) and been predicted for many decades (Smithers et al., 1990). In Type II hybrid design thinking, AI is the sole definer of the overall problem being solved and decider of activities. It is significant that within type II 'hybrid' design thinking, humans are conceived of as resources being called upon by the AI as social, epistemic, and material resources. The ethics of such situations are complex and require deep consideration (Coeckelbergh, 2020; Weisz et al., 2024). This situation is only on the borderline of being feasible with existing technology. AIs have a very different way of structuring knowledge when compared to the human brain, yet the notion of *activities* within a *frame* is still applicable for design agents that use various kinds of concepts.

There are examples of autonomous AI design systems making use of humans that come close to satisfying the criteria for type II hybrid design thinking within limited domains. The ANGLINA system autonomously creates video games and has, in one of its incarnations, involved making requests of a human community to support its artistic/designerly endeavours (Cook et al., 2016; Figure 7.3). This system had clear limitations as an example of type II hybrid design thinking, with many functions being hard-coded and its ability to 'design' rather than create being debatable. The system has served as a basis for more nuanced and ongoing investigation of the possibilities for two-way communication between humans and AIs within the flow of design (McCormack et al., 2020). Across diverse domains researchers are looking towards type II systems, such as in the case of designing therapeutic peptides (Goles et al., 2024).

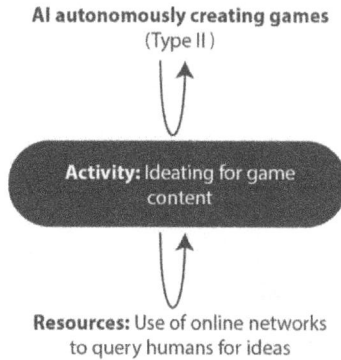

AI autonomously creating games
(Type II)

Activity: Ideating for game content

Resources: Use of online networks to query humans for ideas

FIGURE 7.3 An AI system designing games draws on humans as a resource for ideas through online networks as an example of type II hybrid design thinking.

7.3.3 Type III: Hybrid Problem Framing as Design Team, Mixed Activities

In type III hybrid design thinking, each of the human(s) and AI(s) have their own framing of the design problem and collaborate in a way that is analogous to a human design team with one or more members being AIs – we will refer to all members of the team as 'agents'. Each agent has their own internal framing for the design task, yet they influence each other's framing through a process of externalisation and interpretation (Gero & Kannengiesser, 2007). Any agent within a design team can designate or carry out design activities, either alone or as a collaborative exercise. For example, a brainstorming session is an activity that might be carried out synchronously by all members of a design team with the goal of each agent externalising their framing of the problem and thus both *expanding* and *aligning* everybody's frame. The distinction between type I and type III hybrid design thinking is that in type III the AI-as-agent has its own conception of the design problem and is doing its own work to create a useful assemblage of concepts within which a desired design response might be found.

Similar to type II, there are no examples presently available that fully demonstrate what type III hybrid design thinking looks like. Several interactive co-creative design systems published to date are approaching type III capabilities, meaning that the AI takes on some part of the creative responsibility of framing, if not yet the full capacity to frame and act that the human design(s) possess. One example of such a system is Reframer (Lawton, Grace, et al., 2023; Lawton, Ibarrola, et al., 2023), a collaborative sketching AI in which a human draws with a stylus on a graphics tablet while – in real time – an AI manipulates both human-drawn and AI-drawn vector elements. This differs from typical (and type I) text-to-image AI systems in which the human has little or no input after a prompt is provided. A human-AI co-drawn image of Reframer's interface can be seen in Figure 7.4. In qualitative studies of Reframer, users recalled moments where new visual elements 'emerged' from the lines being moved by the system, causing them to re-frame their drawing. The authors propose a model in which mixed-initiative

FIGURE 7.4 Reframer as a system looking towards AI systems that support type III hybrid design thinking.

design systems like Reframer move between being tool-like support interfaces to being agent-like co-creative partners (what we would call being a type III framing system) as users extend and withdraw agency to/from it while creating. Reframer's capacity for the human and AI to mutually influence each other's framing suggests how future type III systems might change how we design.

Two other examples that look towards type III hybrid design thinking are the Creative Sketching Apprentice (CSA) (Karimi et al., 2019) and LuminAI (Long et al., 2017). In the CSA, a human begins sketching while an AI tries to detect what they are drawing and then propose a concept that is radically different, but visually similar. These intentionally obtuse reframings, which the authors call 'conceptual shifts', are accompanied by sketches showing strong visual similarity between the two concepts, intending to provoke the human into adopting the analogy and further reframe their concept, schematised in Figure 7.5 as a

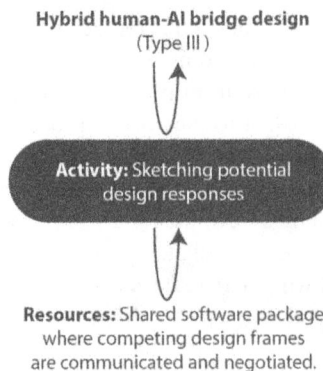

FIGURE 7.5 A hybrid design team are designing a bridge and using a tool (inspired by CSA) that allows them to influence each other's design framing through shared sketch-and-response, as an example of type III hybrid design thinking.

potential example of type III hybrid design thinking. LuminAI is a collaborative improvisational human-AI dance platform that sits in the domain of art
rather than design but is useful for showing how AIs can interact with humans
in an embodied way. Participants in a dome-shaped installation with projections
on its inner surface are matched with AI 'silhouettes', which improvise dance
moves in response to their human partners' movements. The dance moves are
AI-generated in response to those of the human but can both 'follow' the human
(repeating their movements with little adaptation) as well as 'lead' them (generating new movements and evaluating whether the human follows). This is a
poetic analogy for the way that Type III systems can work in future, both leading
and following.

7.4 DISCUSSION

7.4.1 A FRAMEWORK FOR AI IN DESIGN THINKING

The adoption of AI within design differs from the adoption of prior technologies. With AI, there is a future in which (for better or for worse) humans have the
option of delegating the core designerly work of framing design problems and
designating activities. There is now a spectre of problem-owners drawing upon
AIs as designers who in turn draw upon human resources in the scope of design
activities. There is a need for this kind of theory-building to make sense of key
distinctions during these developments, alongside related work considering the
ethics of these different types of hybrid design thinking and the evolving implications for design professionals.

The framework provides a useful lens for revisiting existing conceptions of AI
in design. For example, Weisz et al. (2024) identify three useful, new heuristics
for designers working with generative AI: (1) designing for generative variabil-
ity, helping users to manage the range of outputs from a system; (2) design for
co-creation between people and AIs, helping users to work in partnership with
AIs; and (3) design for imperfection, helping users to understand and work with
unexpected outcomes from generation. These heuristics have different implications depending upon the type of design thinking being considered: a generative
AI system to aid in type I designing (aiding humans in design framing) should
behave differently to a system doing type II designing (AI design framing drawing on humans as a resource).

The framework can also be of assistance in studies considering the use of generative AI tools in design thinking. In their study of OpenAI's GPT-3.5 applied
to different design thinking activities, Fischer et al. (2023) conclude that "the
chatbot does solely augment human capabilities and is not capable of replacing
human design thinkers and human-generated insight" (p. 163). The implication of
the framework for AI use in design thinking – and of the frame-activity model of
design thinking – is that we can be far more precise in the claims being made in
such studies and move beyond discussion of 'augmentation'. Their claims could
be rewritten in a more precise fashion to suggest that there is demonstrable evidence that GPT-3.5 has utility for type I design thinking with AI – and they list

specified types of design activities within which they have found it useful to different degrees – yet they also found that this system is not useful for type II or type III design thinking with AI.

7.4.2 Critical Analysis of AI in Design Thinking

Technologies can be used for good or ill within the limits of their affordances. AI has such broad affordances – especially considering that the pace of AI development has been exponential in recent years – that it is difficult to forecast possible triumphs and pitfalls from societies adopting AI in design thinking. Despite this, it is important to reiterate some of the emergent concerns around AI in design thinking to maintain a critical perspective.

First, there is potential for design jobs to be replaced by AI systems in the case of type II systems or for the number of professional designers required by a society to shrink through type I or type III systems. Given the remoteness of type II systems, critiques have tended to focus upon ways that we can ensure that type I and type III systems are as prosocial as possible (Hoque, 2024).

Second, there is concern that where AI is used in design thinking, it can have a limiting effect upon human creativity. For example, designers believing that they are conducting design research, ideating, and testing by drawing on the resources of AI systems while being unaware of hidden constraints. AI systems rely upon a training set of data and interpolation; this leads to a conceptual space that is necessarily constrained compared to the open world. Further, if we take a broad perspective of designers around the world working on similar problems and take as a given that diversity of design responses is a good thing for the culture of design, then a high reliance upon AI across the profession can lead to a lack of diversity and a form of collective fixation.

Third, and relatedly, the training sets that are used by AI systems are biased in different ways that tend to reflect the dataset being used but not the world outside of it (Ntoutsi et al., 2020). The use of AI by designers can perpetuate bias, especially given the growing practice of using AI when conducting design research and user testing.

The proposed framework has utility for considering each of these three points. Type I systems, which we already see in the world, have a particular set of implications; they tend to be focused upon ensuring that the practice of design continues to be a "dialectical process between tradition and transcendence" (Ehn, 1988, p. 66). From this view, a synthesis of the new technology of AI with continuing traditions will (through judicious usage) lead to improvements in design practice. In contrast, type II systems (and perhaps type III systems) represent a radical change in the role of design. Through design, humans shape their world to suit their preference. If this work of *imagining what preferred situations look like* were to be delegated, then design philosophy writ large would require radical new foundations.

7.5 CONCLUSION

This chapter has outlined a frame-activity model of design thinking by drawing on the design cognition literature. It has developed a framework based on this model that is claimed as useful for analysing human and AI collaboration in design thinking. It has specifically distinguished three types of hybrid design thinking that are characterised by who is doing the design framing: humans, AIs, or the two together. The main contribution of the framework is to propose an alternative to descriptions of design thinking that rest upon analyses of activities and roles within them.

The framework can also be used to analyse the current research agenda for AI in design. Currently, there is much focus upon type I systems that can serve human designers by providing additional, previously unavailable resources within design activities. There are warranted ethical concerns around the prospect of type II systems that draw upon humans or that carry out autonomous design thinking. There is significant scope for more work to be carried out in developing AI systems that are capable of participating in type III hybrid design thinking. Systems that co-operate with humans as co-designers rather than as a resource through an ability to frame design problems and improve that frame through design activities. Such systems are likely to be domain-specific (e.g., the sketching examples discussed); there is no need for artificial general intelligences to make a significant contribution. We hypothesise that such systems might provide a valuable new way to improve the functioning of design teams with utility to shift the framing of human co-designers into directions that might otherwise not be available.

NOTES

1 Concept here refers to units of knowledge within a cognition. We follow a situated notion of concepts in this description, where concepts for human designers are understood to be embodied, may involve actions, and are entangled with emotions, social relationships, and environment: (Barsalou, 2005; Clancey, 1997; Kelly & Gero, 2014, 2021).
2 Or, alternatively, both concept and knowledge space (Hatchuel & Weil, 2003), but we will avoid discussion of C-K theory as this theory uses the term "concept" in a heterodox way and will add unnecessary ambiguity for many readers.
3 The nine-dot problem asks a solver to join a 3x3 grid of dots with three straight and contiguous lines. This problem cannot be solved until the solver realises that they are able to go outside of the implied bounds of the grid. Once this implicit constraint is recognised and overcome, the problem is easily solved.
4 Within the discourse of activity theory, an activity would be described as composed of actions and operations, distinguished by a decreasing importance of intention at each layer in this hierarchy. Such an account sits outside the scope of this chapter, though it is entirely compatible with the framework proposed here.

REFERENCES

Barsalou, L. W. (2005). Abstraction as dynamic interpretation in perceptual symbol systems. In *Building object categories in developmental time* (pp. 407–450). Psychology Press.
Berkun, S. (2020). *How design makes the world*. Berkun Media LLC.

Bölek, B., Tutal, O., & Özbaşaran, H. (2023). A systematic review on artificial intelligence applications in architecture. *Journal of Design for Resilience in Architecture and Planning*, *4*(1), 91–104. https://doi.org/10.47818/DRArch.2023.v4i1085

Cash, P. J. (2018). Developing theory-driven design research. *Design Studies*, *56*, 84–119. https://doi.org/10.1016/j.destud.2018.03.002

Clancey, W. J. (1997). *Situated cognition: On human knowledge and computer representations*. Cambridge university press.

Coeckelbergh, M. (2020). *AI ethics*. MIT Press.

Cook, M., Colton, S., & Gow, J. (2016). The Angelina videogame design system: Part I. *IEEE Transactions on Computational Intelligence and AI in Games*, *9*(2), 192–203. https://doi.org/10.1109/TCIAIG.2016.2520256

Coyne, R., Rosenman, M., Radford, A., Balachandran, M., & Gero, J. (1990). *Knowledge-based design systems*. Addison Wesley.

Cross, N. (2023). Design thinking: What just happened? *Design Studies*, *86*. https://doi.org/10.1016/j.destud.2023.101187

Dong, A., Kleinsmann, M. S., & Deken, F. (2013). Investigating design cognition in the construction and enactment of team mental models. *Design Studies*, *34*(1), 1–33. https://doi.org/10.1016/j.destud.2012.05.003

Dorst, K. (2011). The core of 'design thinking' and its application. *Design Studies*, *32*(6), 521–532. https://doi.org/10.1016/j.destud.2011.07.006

Dorst, K. (2015). *Frame innovation: Create new thinking by design*. MIT press.

Dorst, K., & Cross, N. (2001). Creativity in the design process: Co-evolution of problem–solution. *Design Studies*, *22*(5), 425–437. https://doi.org/10.1016/S0142-694X(01)00009-6

Ehn, P. (1988). *Work-oriented design of computer artifacts*.

Fischer, H., Dres, M., & Seidenstricker, S. (2023). Application of ChatGPT in design thinking. *Application of Emerging Technologies*, *115*(115). https://doi.org/10.54941/ahfe1004312

Gero, J. S., & Kannengiesser, U. (2007). An ontology of situated design teams. *AI EDAM*, *21*(3), 295–308. https://doi.org/10.1017/S0890060407000297

Goles, M., Daza, A., Cabas-Mora, G., Sarmiento-Varón, L., Sepúlveda-Yañez, J., Anvari-Kazemabad, H., Davari, M. D., Uribe-Paredes, R., Olivera-Nappa, Á, & Navarrete, M. A. (2024). Peptide-based drug discovery through artificial intelligence: Towards an autonomous design of therapeutic peptides. *Briefings in Bioinformatics*, *25*(4). https://doi.org/10.1093/bib/bbae275

Goodyear, P., & Carvalho, L. (2014). Framing the analysis of learning network architectures. In *The architecture of productive learning networks* (pp. 66–88). Routledge.

Hatchuel, A., & Weil, B. (2003). *A new approach of innovative design: An introduction to CK theory*. DS 31: Proceedings of ICED 03, the 14th International Conference on Engineering Design, Stockholm.

Hoque, F. (2024). Does artificial intelligence have the possibility of taking over Designers' jobs in the future. *International Journal of Science and Business*, *31*(1), 26–35. https://doi.org/10.58970/IJSB.2273

Johansson-Sköldberg, U., Woodilla, J., & Çetinkaya, M. (2013). Design thinking: Past, present and possible futures. *Creativity and Innovation Management*, *22*(2), 121–146. https://doi.org/10.1111/caim.12023

Karimi, P., Grace, K., Davis, N., & Maher, M. L. (2019). Creative sketching apprentice: Supporting conceptual shifts in sketch ideation. In J. S. Gero (Ed.), *Design computing and cognition '18* (pp. 721–738). Springer International Publishing. https://doi.org/10.1007/978-3-030-05363-5_39

Karimi, P., Grace, K., Maher, M. L., & Davis, N. (2018). Evaluating creativity in computational co-creative systems. *arXiv Preprint arXiv:1807.09886.*

Kelly, N., & Gero, J. S. (2014). Interpretation in design: Modelling how the situation changes during design activity. *Research in Engineering Design, 25,* 109–124. https://doi.org/10.1007/s00163-013-0168-y

Kelly, N., & Gero, J. S. (2021). Design thinking and computational thinking: A dual process model for addressing design problems. *Design Science, 7,* e8. https://doi.org/10.1017/dsj.2021.7

Kelly, N., & Gero, J. S. (2022). Reviewing the concept of design frames towards a cognitive model. *Design Science, 8,* e30. https://doi.org/10.1017/dsj.2022.25

Kerr, J., Cheers, J., Gallegos, D., Blackler, A., & Kelly, N. (2023). *The art of co-design: A guide to creative collaboration.* BIS Publishers.

Kimbell, L. (2011). Rethinking design thinking: Part I. *Design and Culture, 3*(3), 285–306. https://doi.org/10.2752/175470811X13071166525216

Lawton, T., Grace, K., & Ibarrola, F. J. (2023). *When is a tool a tool? User perceptions of system agency in human–ai co-creative drawing.* 1978–1996.

Lawton, T., Ibarrola, F. J., Ventura, D., & Grace, K. (2023). Drawing with Reframer: Emergence and Control in Co-Creative AI. *Proceedings of the 28th International Conference on Intelligent User Interfaces,* 264–277. https://doi.org/10.1145/3581641.3584095

Long, D., Jacob, M., Davis, N., & Magerko, B. (2017). Designing for Socially Interactive Systems. *Proceedings of the 2017 ACM SIGCHI Conference on Creativity and Cognition,* 39–50. https://doi.org/10.1145/3059454.3059479

Maher, M. L., & Poon, J. (1996). Modeling design exploration as co-evolution. *Computer-Aided Civil and Infrastructure Engineering, 11*(3), 195–209. https://doi.org/10.1111/j.1467-8667.1996.tb00323.x

McCormack, J., Hutchings, P., Gifford, T., Yee-King, M., Llano, M. T., & D'inverno, M. (2020). Design considerations for real-time collaboration with creative artificial intelligence. *Organised Sound, 25*(1), 41–52. https://doi.org/10.1017/S1355771819000451

Murphy, G. (2004). *The big book of concepts.* MIT press.

Nardi, B. A. (1996). *Context and consciousness: Activity theory and human-computer interaction.* MIT Press.

Ntoutsi, E., Fafalios, P., Gadiraju, U., Iosifidis, V., Nejdl, W., Vidal, M., Ruggieri, S., Turini, F., Papadopoulos, S., & Krasanakis, E. (2020). Bias in data-driven artificial intelligence systems—An introductory survey. *Wiley Interdisciplinary Reviews: Data Mining and Knowledge Discovery, 10*(3), e1356. https://doi.org/10.1002/widm.1356

Pena, M. L. C., Carballal, A., Rodríguez-Fernández, N., Santos, I., & Romero, J. (2021). Artificial intelligence applied to conceptual design. A review of its use in architecture. *Automation In Construction, 124,* 103550.

Ploennigs, J., & Berger, M. (2023). AI art in architecture. *AI in Civil Engineering, 2*(1), 8. https://doi.org/10.1007/s43503-023-00018-y

Schmidt, A., Elagroudy, P., Draxler, F., Kreuter, F., & Welsch, R. (2024). Simulating the human in HCD with ChatGPT: Redesigning interaction design with AI. *Interactions, 31*(1), 24–31. https://doi.org/10.1145/3637436

Schon, D. A. (1983). *The Reflective Practitioner. How Professionals Think in Action.*

Schön, D. A., & Rein, M. (1994). *Frame reflection: Toward the resolution of intractable policy controversies. Basic Book.*

Seidel, S., Berente, N., Lindberg, A., Lyytinen, K., Martinez, B., & Nickerson, J. V. (2020). Artificial intelligence and video game creation: A framework for the new logic of autonomous design. *Journal of Digital Social Research, 2*(3), 126–157. https://doi.org/10.33621/jdsr.v2i3.46

Smithers, T., Conkie, A., Doheny, J., Logan, B., Millington, K., & Tang, M. X. (1990). Design as intelligent behaviour: An AI in design research programme. *Artificial Intelligence in Engineering*, *5*(2), 78–109. https://doi.org/10.1016/0954-1810(90)90004-N

Sosa, R., Connor, A. M., & Corson, B. (2017). Framing creative problems. In *Handbook of research on creative problem-solving skill development in higher education* (pp. 472–493). IGI Global.

Sreenivasan, A., & Suresh, M. (2024). Design thinking and artificial intelligence: A systematic literature review exploring synergies. *International Journal of Innovation Studies*. https://doi.org/10.1016/j.ijis.2024.05.001

Suwa, M., Gero, J., & Purcell, T. (2000). Unexpected discoveries and s-invention of design requirements: Important vehicles for a design process. *Design Studies*, *21*(6), 539–567. https://doi.org/10.1016/S0142-694X(99)00034-4

Valkenburg, R., & Dorst, K. (1998). The reflective practice of design teams. *Design Studies*, *19*(3), 249–271. https://doi.org/10.1016/S0142-694X(98)00011-8

Weisz, J. D., He, J., Muller, M., Hoefer, G., Miles, R., & Geyer, W. (2024). Design Principles for Generative AI Applications. *Proceedings of the CHI Conference on Human Factors in Computing Systems*, 1–22. https://doi.org/10.1145/3613904.3642466

Wendrich, R. E. (2013). *Hybrid Design Thinking in a Consummate Marriage of People and Technology*. 5th World Conference on Design Research, IASDR 2013: Consilience and innovation in design.

Yüksel, N., Börklü, H. R., Sezer, H. K., & Canyurt, O. E. (2023). Review of artificial intelligence applications in engineering design perspective. *Engineering Applications of Artificial Intelligence*, *118*, 105697. https://doi.org/10.1016/j.engappai.2022.105697

Zahedi, M., Tessier, V., & Hawey, D. (2017). Understanding collaborative design through activity theory. *The Design Journal*, *20*(sup1), S4611–S4620. https://doi.org/10.1080/14606925.2017.1352958

Zelenko, O., Gomez, R., & Kelly, N. (2021). Research co-design: Meaningful collaboration in research. In *How to be a design academic* (pp. 227–244). CRC Press.

8 High-Performance Flight Vehicle Design

*Glenn Gebert, Eric Stenftenagel,
and Nick Blanton*

8.1 INTRODUCTION: SYSTEM-LEVEL DESIGN AND OPTIMISATION

As with the general application of design thinking to complex systems, the design of virtually all flight vehicles involves the interaction of multiple disciplines and competing constraints. An aircraft needs to be as light as possible to extend its range, yet strong enough to withstand its most extreme loading. The aircraft must carry the largest amount of fuel, but still have sufficient volume for its cargo and instrumentation. In addition, propulsion systems must meet desired flight and manoeuvring parameters but still fit within weight and packaging limits. Historically, complex projects necessitate teams of people working on aerodynamics, structures, packaging, propulsion, flight controls, cost, and more. Collaboration occurs between disciplines, but the flow of information between groups of people can at times be slow, incomplete, not understood, or with error. The exchange of information between the parties has largely been a manual process. Furthermore, such segregation of the work does not lend itself to the overall, system-level, optimisation of the vehicle, and, of course, design thinking works to not just generate a design, but endeavours to create the best design. It is necessary to explore a wide design space, include the needs and constraints of all associated disciplines, and quantify performance to identify the leading candidates.

The designing of modern weapon systems introduces additional constraints that are different from conventional aircraft. As weapons are mass produced on an economic scale, the cost must be significantly less than the intended target. Volume limitations associated with the carriage and launch environment often necessitate the deployment of surfaces after release. The launch environment from carriage aircraft can lead to high angular rates, incidence angles, and loading that can drive the structural maximum of the surfaces. Furthermore, end-game manoeuvring loads can far exceed the maximum associated with human controlled aircraft. Even though this situation may exist for only a few seconds, the vehicle must be designed to be controllable and sufficiently stable to endure the rest of the mission.

In recent years, design thinking has emerged as a valuable approach for addressing these challenges and facilitating multidisciplinary design as discussed by Plattner, Meinel, and Leifer [1]. The problem-solving approach associated with

DOI: 10.1201/9781003487524-8

design thinking emphasises analysis, creativity, detailed modelling, and performance assessment. It involves a series of iterative steps, as the performance of potential designs are compared to others and evaluated against other solutions that may have merit. Solutions are updated and refined by enhancing modifications that improve performance and reducing detrimental changes. Complex systems involve the tight coupling of multiple disciplines, which often have competing requirements. The design thinking process requires the designer to consider the impact of each sub-system, other disciplines and the overall performance. Although such interconnectivity can be challenging, modern computer throughput and learning algorithms allow modelling of multiple systems and address their causal relationships. The results allow the designer to stay up to date with detailed information on the impact of each discipline on the other and the assessment of the total performance against each of the design goals. This approach is well-suited to the multidisciplinary design of weapon systems, as it allows for the integration of fields, correlates total system results, and provides the designer with Pareto-front performance data for each of the design goals so that the best solution can be identified.

The use of the automated multidisciplinary design and optimisation (MDO) process has the potential to enhance the design thinking process, since an extremely wide range of solutions can initially be evaluated allowing what is typically regarded as "out-of-the-box" concepts being modelled and evaluated as part of the total dataset. Original MDO work can be traced to Schmit, Haftka et al. [2–9] coupling their structural analyses to other disciplines. Much later, the MDO concept was extended to the design of aeroelastic wings [10–17] and complete aircraft [18–22]. An excellent summary of the MDO process and its history is provided by Martins and Lambe [23]. Through the MDO procedure, initial promising solutions are first identified, and the designer can narrow the analysis space to focus on the most favourable designs, and the process is repeated as its refinement continues. By initially exploring a very large design space, the engineer is afforded the opportunity to examine the performance of atypical options. The wide design space exploration enhances creativity and experimentation in the design process. This is important in the context of weapon systems, where the stakes are high, and the challenges are complex. By allowing designers to think outside the box and to experiment with different ideas, the MDO process facilitates design thinking. After the broad understanding of the design goals are determined, the MDO process identifies the critical inputs and quantifies critical outputs which drive the design. The learning algorithms, automated information flow, feedback of performance results, and wide design space exploration allow the engineers to foster innovation and to identify novel solutions to difficult problems. As solutions are identified, the results are returned into the broader design thinking process and used to refine the overall design objectives, if necessary.

Typically, a key aspect of design thinking is its iterative nature. Rather than following a linear process, it involves a series of cycles of understanding, generating, testing, and refining. This often allows designers to learn from their mistakes and continuously improve their designs based on feedback and new insights.

Overall MDO Process

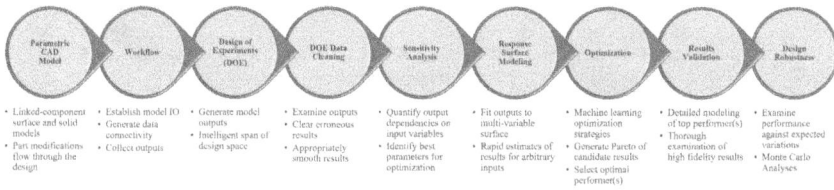

FIGURE 8.1 Multidisciplinary design and optimisation process.

While this has generally been a labour-intensive process, the integration of critical modelling tools allows for the automation of these iterations as the designs converge, and the engineer can evaluate more mature solutions and weigh their performance against the competing design goals. The MDO process determines each vehicle's performance against the requirements and identifies the optimal. This process is illustrated in Figure 8.1, with each component being parametric Computer-Aided Design (CAD) model, workflow, design of experiments (DOE), DOE data cleaning, response surface modelling, sensitivity analysis, optimisation, results validation, design robustness. The MDO process allows for a rapid examination of the entire range of the design space to identify the best conceptual design to address the requirements. Once the first-order design is determined, the parameter space can be narrowed to refine the results with increasingly higher-fidelity models. The following sections describe each step in detail.

8.1.1 PARAMETRIC CAD MODEL

Design thinking is a problem-solving approach that focuses on the importance of a complete and interconnected view of the system being designed. This philosophy is highly relevant to the concept of parametric CAD models within the MDO process, where identifying and automating the interconnections between various geometric components of a design allows for the rapid examination of a small number of critical parameters. During the conceptual design phase, a "single source of truth", parametric, CAD model facilitates rapid prototyping and concept iteration. It is important that the CAD is created with a sufficiently large range of variables to encompass the widest reasonable design space since the parameters will limit the overall set of examined designs. With only a small number of changes to the input, the parametric CAD model can automatically update to new designs and provide essential information for downstream tools and simulations, demonstrating the power of this interconnected approach to problem-solving. The parameterized CAD model contains scripted interconnections linking all the geometric components of the design to a small number of inputs. A change in a body diameter not only changes the frontal area of the fuselage, but it also affects the location of the wing and tails, the fineness ratio of the nose, the wing/body interface, size of fuel tanks, location of avionics, and much more. Similar interconnectivity

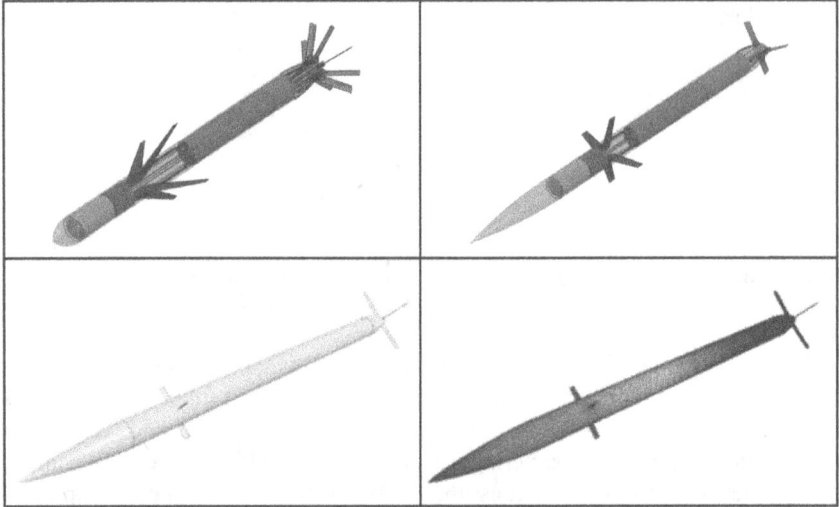

FIGURE 8.2 Representation of parametric CAD leading to rapid design change and associated CFD grids.

is associated with many of the geometric values of a design. It is critical to identify and script these interconnections, and make the entire geometry specified by a small number of critical parameters. Parametric CAD model robustness must also be achieved to ensure all planned combinations of input parameters generate valid vehicle geometries. The parametric CAD model also leads to the examination of a small number of critical parameters with automatic regeneration of CAD files associated with each modification. These CAD files can then be imported into a computational fluid dynamics (CFD) grid generation package. Most CFD grid generators automatically generate a mesh based on a small number of preselected settings (i.e. base mesh size, number of surface cells, prism layer thickness, etc.). The top two images in Figure 8.2 illustrate different designs, which alter the number and location of canards and tails as well as vehicle length, diameter, and nose bluntness. The lower two images show the CAD for a typical vehicle and the associated CFD computational domain. With only a small number of changes to the inputs (number and location of surfaces, body diameter, nose bluntness, etc.), the parametric CAD updated to the new designs, automatically imported the geometry into the grid generator, and the new CFD results were created.

8.1.2 Workflow

Workflow plays a critical role in assisting designers by automating the analysis of a particular design. By facilitating problem-solving, systems thinking, iteration, and user-centred decision-making, the workflow enables designers to explore a wide design space, identify critical measures of merit, and assess the performance

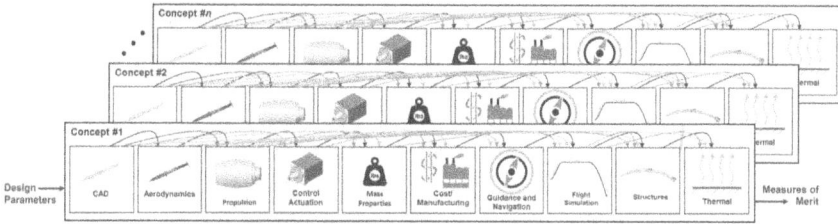

FIGURE 8.3 Typical modules and information flow within a flight vehicle optimisation workflow.

of candidate designs, ultimately leading to the development of an optimal solution. Every design has parameter limits, a set of goals, and every associated discipline has its particular requirements. Air vehicles often desire to maximise range, minimise fuel consumption, increase their payload, decrease their cost, minimise their size, etc. Reducing fuel capacity and consumption will lower the vehicle size, but this is at odds with increasing the range. Exotic, lightweight materials can decrease the bulk, but may increase manufacturing costs, and likewise, components such as high bandwidth control actuators will have high power requirements. A workflow is therefore created to determine the sufficient basis of input parameters to explore a wide design space and identify the critical measures of merit associated with the desired optimal design.

As shown in Figure 8.3, the flow of information between separate modules is first established. Each block receives all the information of the design to produce its estimations, and only the information that is required. Each module then outputs its required information to all other appropriate routines. With new combinations of input parameters, the workflow generates candidate vehicles including the Outer Mold Line (OML), critical components, associated internal layouts, etc. The associated aerodynamics, mass properties, manufacturing costs, etc. can then be calculated. Once all units are determined, a flight simulation is executed to evaluate the performance. This process is performed for *n* concepts, and the measures of merit for each design are determined.

8.1.3 DESIGN OF EXPERIMENTS

DOE is a valuable tool in the MDO process that intelligently investigates all concepts within the design space to determine appropriate trends. By using the DOE approach, the MDO process can focus on a small subset of configurations within the design space, providing sufficient data to predict the performance of any desired configuration. The use of DOE in the MDO process allows for the inspection of various design options and provides the ability to pivot and adjust as needed.

Most vehicles require a reasonably large number of parameters to define. For example, a simple typical airframe requires wing and tail locations, root and tip chords, surface span, number and orientation of tail surfaces, type of propulsion,

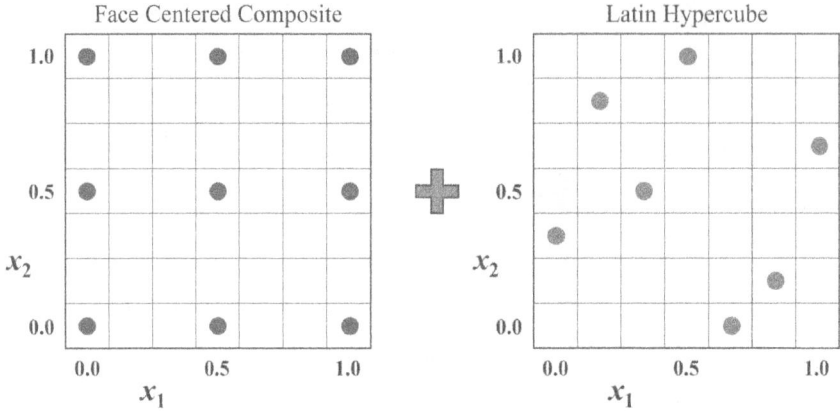

FIGURE 8.4 Typical DOE design space sampling.

fuselage/body dimensions, control surface actuation methods, and more. Rather than examining the full matrix of a multidimensional design, a DOE approach intelligently examines critical concepts within the design space to determine the appropriate trends. With this smaller subset of data, the coverage of the DOE is often sufficient to allow for the estimated performance of any candidate vehicle within the entire design space. Typically, results across the complete design space are examined using a face-centred central composite and Latin hypercube design point distributions or some similar techniques [24]. As in the simplified model shown in Figure 8.4, the Latin square selects parameter with one and only one value in each row and column. This Latin hypercube extends its concept to an arbitrary number of dimensions. While the Latin hypercube covers the design space, it would often require extrapolations to the corner and edge points. The face-centred central composite guarantees that all extreme values are considered and only interpolations, rather than extrapolations are required.

Continuing the example stated earlier, the DOE results can provide all aerodynamic coefficients for most essential parameters (wing and tail locations, root and tip chords, etc.). These results are then available for interpolation to any geometry within the design space. It is important to note that DOE applies to all the modules associated with the workflow. Initial estimates used in the DOE are typically generated via lower order models. As such, there are uncertainties in the results that must be considered throughout the design process.

8.1.4 DOE DATA CLEANING

Lower-order models used in the DOE may lead to errors in the values and trends. The raw data may sometimes show considerable scatter and errant results. The data is employed to construct response surface models (RSM) and is used to interpolate the parameters associated with arbitrary inputs within the domain.

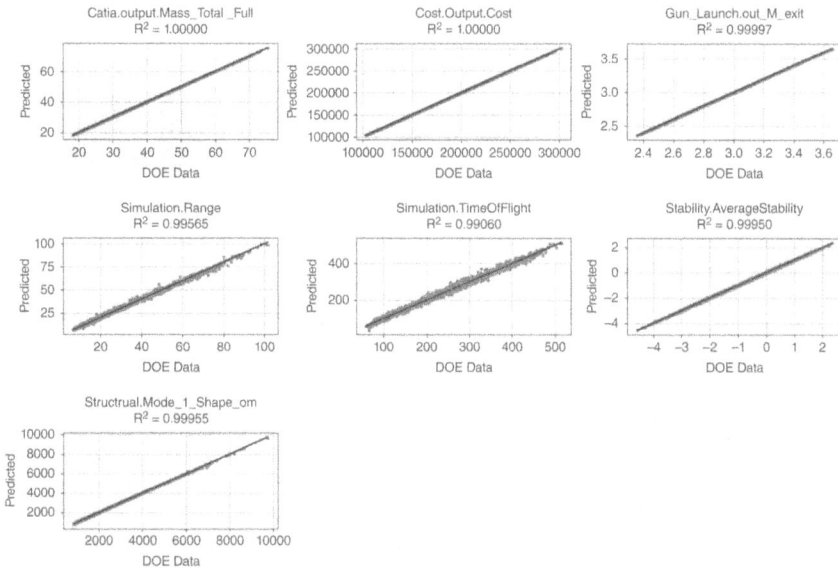

FIGURE 8.5 Errant data resulting from the machine learning algorithms.

Inaccurate, erratic results lead to poor quality response surfaces and errored interpolations. As such, results must be intelligently smoothed to maintain the pertinent trends but reduce computational noise. The typical machine learning techniques used to construct RSM will effectively create hypersurfaces to reduce "jitters" in the data, but errant values can still occur as sparce data points that overshoot. After sufficient DOE data cleaning, an example high quality RSM is shown in Figure 8.5. Techniques and algorithms are used to identify and adjust these outliers, but manual evaluation of the RSM fits is typically required to verify the smoothness and output range.

8.1.5 RESPONSE SURFACE MODELLING

RSM facilitates design thinking by enabling visualisation, identifying critical input variables, exploring the design space, rapidly evaluating candidate designs, and encouraging human evaluation. By incorporating RSM into the design process, designers can make more informed decisions, alter the design space, and ultimately create more innovative and effective solutions. The cleaned data generated from the DOE are used to construct multi-dimensional RSM results. The ultimate goal of RSM is to determine the best-fit hyper-curve through the data. A variety of options are available for the generation of these models. Supervised learning regression models (SLRM), such as H2O's Deep Learning[1] employs a feed-forward artificial neural net. Typically, the neural net is trained with a stochastic gradient descent using back-propagation. RSM that is generated from the DOE is often in the form of tabular data. Therefore, a deep neural network (DNN)

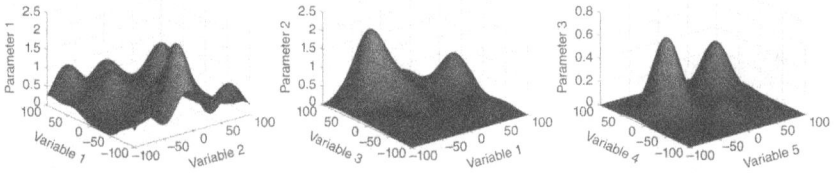

FIGURE 8.6   Typical response surface models.

or multilayer perceptron (MLP) is used to provide a richer understanding of the data. Most RSM generate the output from many input variables. As such, they are difficult to visualise. Human evaluation of the RSM requires fixing several of the input variables to observe the RSM's behaviour with respect to several other variables. A typical set of RSM is shown in Figure 8.6.

The RSM is constructed from a relatively small number of intelligently selected data points resulting from the DOE distribution strategy selected and can sometimes miss important non-linearities and discontinuities. The RSM results must be evaluated for rationality. Once created, RSM results allow for rapid predictions through interpolation. It is important to note that a well-constructed set of RSM allows for the rapid evaluation of all candidate design options. Rather than solving a complex CFD, structural and cost model for each configuration, the estimated results can be quickly determined through interpolation within the RSM.

8.1.6   SENSITIVITY ANALYSIS

Evaluating the RSM results provides valuable insights into which parameters generate the largest contributions to the desired vehicle's behaviour. Design thinking principles can be applied in this context by initially focusing on parameters that provide the most benefit to the desired performance. As the design evolves, parameters with a weaker effect on the design can be more thoroughly scrutinised, allowing designers to refine their understanding of the design space and to make informed decisions. The RSM approach quantifies the results for any set of input parameters within the evaluation domain. For a flight vehicle, the aerodynamics, structural properties, cost, etc. can all be rapidly estimated through quick interpolation of the RSM. The natural frequency of the wing, the lift and drag, the weight, the specific fuel consumption, and more are rapidly determined from changes in the wingspan, engine placement, nose fineness, etc. As such, evaluation of the RSM results provide information in which those parameters then generate results that represent the vehicle's desired behaviour. A typical schematic of the sensitivity of aerodynamic performance on typical wing input parameters is shown in Figure 8.7.

Broad design optimisation initially focusses on the parameters that provide the most benefit to the desired performance. The parameters having a weaker effect on the design are then more scrutinised as the design evolves. If an input parameter is found to have little or no impact on the desired vehicle behaviour, it could potentially be eliminated as an input variable leading to a simplified RSM.

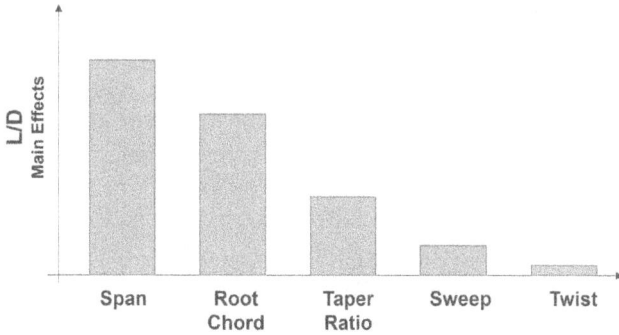

FIGURE 8.7   Typical sensitivity results.

8.1.7 OPTIMISATION

With the workflow and RSM established, the entire design space is examined and optimal configurations are identified. Constraints are assigned, and each candidate is evaluated against its ability to address all the requirements. By employing the surrogate model, designers can rapidly test and evaluate tens of thousands of designs, making the design process more efficient and effective. Multi-discipline optimisation typically has multiple design goals. These goals compete and are optimised for different configurations. As such, the overall optimal design is typically a compromise for each individual goal to achieve the best overall solution. There are many ways to search within the design space and to identify the optimal configuration. One method that has been demonstrated to work well is the non-dominated sorting genetic algorithm II (NSGA-II). The NSGA-II simulates Darwin's theory of natural selection via a genetic algorithm approach. A population of candidate solutions is made, which evolve towards the best solution to solve a multi-objective optimisation. Each design is evaluated against a specified fitness function or set of fitness functions, which account for and appropriately weighed with all desired goals. The NSGA-II model will calculate the Pareto front that corresponds to a set of nondominated solutions, which provide a suitable compromise between all objectives, as shown in Figure 8.8.

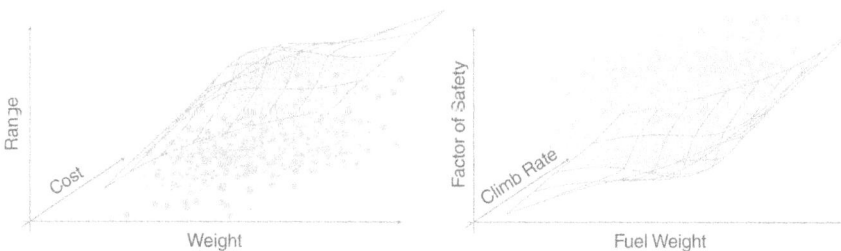

FIGURE 8.8   Typical pareto front of the performance for design options.

8.1.8 Results Validation

To ensure accurate and reliable results, it is important to check the accuracy and completeness of the data, the appropriateness of the statistical models used, and the significance of the results, as well as ensuring that the results are reproducible. It is crucial to thoroughly validate the MDO process and its outcomes to ensure that the design, as a whole, meets the system-level requirements of all disciplines included. Results are carefully scrutinised to verify that the response surfaces are modelling the trade space correctly. In addition, requirements themselves are also scrutinised to ensure that they are reasonable. As designs are examined, it often becomes apparent that the requirements used for the optimisation may sometimes not focus on all the actual vehicle needs. For example, let us consider a vehicle that is being designed for maximum range and minimum gross weight. It must fly at an altitude of at least 18,000 ft and a Mach number range between 0.6 and 0.95. The stowed wings of the cruise missile must deploy for its flyout. As the optimisation occurs, virtually all designs will become a vehicle with a constant wing chord, maximised for volume in the stowed configuration and unswept to minimise their span and subsequent weight. All designs will fly at the minimum altitude of 18,000 ft and a Mach of 0.6. These results achieve the desired goals. However, it is then noted that a high-speed dash at a Mach number of at least 0.85 at sea level is necessary. The designs of the original optimisation cannot achieve such flight conditions, and the optimisation goals or constraints must be altered to include the additional requirement. While it seems possible to determine all requirements before beginning the optimisation to focus on the single best set of solutions, in practice, several iterations are often needed and the design limits and optimisation goals are modified. The insights from initial trials lead to better definition of objectives necessary for the best design.

Following checks, confirming design limits and refinements, the workflow and optimisation will identify a single or a small number of superior performing vehicles being considered. As such, these solutions need to be rapidly determined and the performance is typically estimated with lower order tools. Following the initial design investigation, the design space can be considerably narrowed. With the reduced range of candidate vehicles, higher fidelity tools can be employed to estimate performance. The reduced set of potential airframes are further scrutinised and modelled with higher fidelity tools and techniques to verify their ability to address all of these requirements. High fidelity analysis will focus on not only improving the primary drivers, but also the secondary and tertiary. If required and if time permits, further tools and predictions can occur in the increasingly reduced design space to identify the best design. Figure 8.9 shows how the design space refines as the MDO process resolves.

8.1.9 Design Robustness

The MDO process examines a wide design space and systematically identifies the best overall performer or a small set of vehicles. The process starts

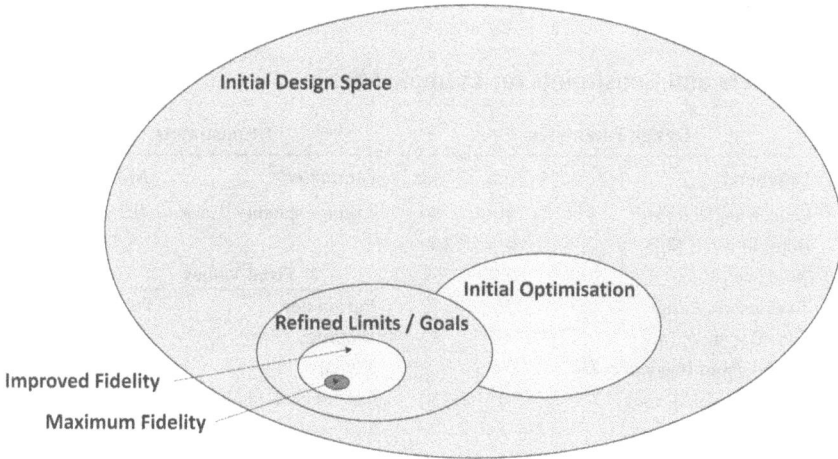

FIGURE 8.9 Reduction in exploration space as design refines.

out with low fidelity modelling and performance predictions. As the design refines, higher fidelity methods are introduced. The final vehicle(s) are those having the best performance against the objective function(s), but total performance has yet been verified. For the final design, detailed aerodynamic, layout, mass, propulsion, flight simulation, cost, etc. models need to be created so that the vehicle's performance in a wide variety of flight scenarios can be evaluated. Various flight and engagement profiles need to be examined as well as atmospheric conditions associated with non-standard atmosphere and winds. In addition, the variation of all appropriate parameters (aerodynamic uncertainties, mass variations, propulsion extremes, etc.) need to be examined either through parameter build-up or a Monte Carlo analysis. The final design is only adequate if it demonstrates robustness in off nominal operating conditions in all flight regimes.

8.2 MULTIDISCIPLINARY DESIGN EXAMPLE

To illustrate the implementation of the MDO framework within a design thinking process, a typical missile design is considered. This workflow can be adapted to the design of any type of flight vehicle. In doing so, a narrow, but sufficient set of parameters are chosen to define the critical geometry, mass properties, propulsion system, lifting surface size and location, and volumetric constraints. For this example, the list of design parameters, fixed values, and constraints are shown in Table 8.1. More parameters could be varied to extend the design space exploration. However, for this example, the number is purposely kept small.

In addition, a set of requirements are established to evaluate the performance of all candidate designs. For this example, the missile is tube launched at a

TABLE 8.1

Parameters and Constraints for Example Missile

Design Parameters			Constraints		
Parameter	**Min**	**Max**	**Parameter**	**Min**	**Max**
Launch angle, θ (deg)	10	80	Average stability (Cal.)	-0.25	0.25
Boost thrust, T (lbf)	100	1000			
Nose fineness, f_n	0.5	4	**Fixed Values**		
Total length, L (in)	36	72	**Parameter**	**Value**	
Canard span, b_c (in)	6	12	Diameter, D (in)	3.75	
Canard chord fraction, c_c/D	0.5	1	Tail taper ratio, r_t	1	
Canard sweep, Λ_c (deg)	0	50	Canard taper ratio, r_c	0.8	
Number of tail fins, N_c	3	6	Tail t_t/c_t	8%	
Tail span, b_t (in)	6	18	Canard t_c/c_c	8%	
Tail deployment angle, Γ_c (deg)	90	140			

variable angle of θ, fires a solid rocket motor to climb and accelerate, reaches apogee, glides downrange, and addresses a fixed ground target. The designs are judged on maximising the range R and minimising the production cost. However, it is possible to include manoeuvring requirements, time of flight, seeker performance, and more. A schematic of the flight and vehicle design is shown in Figure 8.10.

8.2.1 EXAMPLE WORKFLOW OVERVIEW

For each candidate design, an update to independent variables is fed to the workflow and each tool in the workflow is updated with results for the new configuration. Figure 8.11 represents the overall workflow for an example of a missile design.

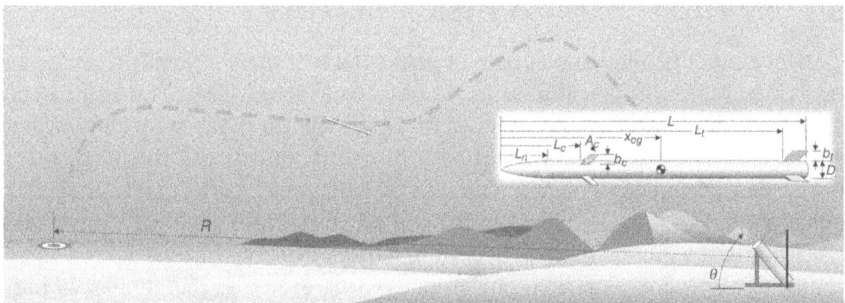

FIGURE 8.10 Schematic of missile design and flight trajectory.

FIGURE 8.11 Workflow for example missile design.

8.2.1.1 Parametric CAD Model

A robust CAD model of the example flight vehicle is first generated such that it can update the designed surfaces and mass properties with any changes made to the relevant independent variables of the study. For this workflow, weight, centre of gravity at full propellant weight, centre of gravity at empty propellant weight, and a full detailed CAD model is generated for use in the downstream tools. Some components such as avionics, warhead, and control systems are fixed sizes. However, these components are still repositioned in the model with each new design. As the missile length changes, the solid rocket motor size is the only internal component that increases or decreases in length to accommodate. As tailfins or canards increase in span, the solid rocket motor size is also reduced to ensure that the tailfins and canards have adequate room to stow within the missile enclosure.

8.2.1.2 Aerodynamic Performance

Typically, first-order tools that run in seconds are used up front for aerodynamic estimates. As stated previously, the fidelity of the aerodynamics tool can increase as the design space narrows. For this example, tools such as Missile DATCOM or VORLAX are used. An array of altitude, Mach, angle of attack, and angle of sideslip are run to build a full aerodynamic database.

8.2.1.3 Propulsion

The missile being designed in this example contains a solid rocket propulsion motor. The propulsion tool in the workflow calculates the thrust versus time curves of the parametric solid rocket motor. The workflow allocates a volume for the solid rocket propellant and a specific impulse of the fuel is assigned. As such, the fuel it takes is cylindrical with a setback from the missile diameter to account for the rocket motor casing. Similarly, the total impulse of the propulsion system, the thrust level, and the action time are set by the single length of the fuel parameter. Furthermore, since well-designed propulsion systems require a particular length for the throat, nozzle, and forward close out, the entire length and mass of the propulsion system is determined by the fuel length.

8.2.1.4 Structural

For aerodynamic performance, flight vehicles need to be as light as possible. In addition, they must be sufficiently strong to withstand aeroelastic and manoeuvring loads. Moreover, autopilot controlled, high-performance flight vehicles will typically have a low pass filter in terms of acceleration on the Inertial Measurement Units (IMU) to remove noise and body bending modes. Without the filter, the autopilot and control system will react to body bending, which leads to excess power consumption and errant behaviour. If the frequency limit is too low, the vehicle will be "sluggish" and will not be able to engage the target. Thus, it is necessary to design sufficiently stiff flight vehicles with their first mode natural frequency sufficiently set high, such that when filtered, it still allows for rapid autopilot response.

In the example of the missile design, the first body bending mode is the lowest frequency and it is only necessary that its frequency is above a threshold value. To that end, the mode shapes and frequencies are calculated for a free-free beam with axially variable density, section moments of inertia, and modulus of elasticity. The material properties are built from the top-level material properties of the outer structure and significant internal components. The nose cone is modelled as ceramic, the body is aluminium, the rocket motor casing is steel, and the propellant is rubber. The overall estimate of stiffness and mass properties are also included for the radar, avionics, warhead, control actuation system, canards, fins, and nozzle.

8.2.1.5   Stability and Control

Stability and control authority is checked every iteration of the design being generated. The workflow calculates each design's mass, moments of inertia, and centre of gravity location. The aerodynamic properties associated with each vehicle's overall missile length, canard geometry, and the tailfin geometry is calculated via semi-empirical or CFD methods. The static and dynamics stability and control effectiveness are calculated and assessed against the design's constraints and manoeuvring requirements. Adequate control authority is required to pull required load manoeuvres and trim throughout the design flight envelope.

8.2.1.6   Gun Launch Simulation

The missile is launched from a gun with an initial propellent charge employed to generate the gun ejection and with a small initial velocity. In addition to the length, diameter, and weight of the missile, the model makes the gun diameter the same as the diameter of the missile in the surface stowed configuration. The length of the charge, length of gap between the charge and the missile, and the total length of the gun barrel are also variables that need to be taken into account. A single degree-of-freedom model moves the missile through the tube accounting for gravity and the push from the propelling charge. The gun launch simulation ends when the aft end of the missile exits the tube.

8.2.1.7   Two Degree-of-Freedom Simulation

A simplified two degree-of-freedom (2DOF) simulation was created to estimate the flight performance of a given configuration. Three phases of flight are modelled including climb, glide, and terminal. During the climb phase, a zero angle of attack body orientation is maintained to execute a gravity turn from the launch flight path angle to apogee. Post apogee, during the glide phase, a max lift-to-drag attitude is maintained to maximise range. The glide phase continues until the vehicle descends below 2,000 ft altitude. At that time, it switches to terminal guidance. During the terminal phase of flight, the angle of attack is commanded back to zero to orient the body frame with the velocity frame for impact.

8.2.1.8 Thermal

Supersonic and hypersonic vehicles can experience significant heating associated with the isentropic compression of the incoming air and the viscosity within the boundary layer. Various components along the vehicle need to be monitored for their peak temperature and heat flux to verify that the design will not weaken or fail. For the example of a missile design, only the heat transfer on the outer skin of the vehicle is modelled using typical external correlations for the convective heat transfer associated with the incoming air, insulated internal surface, since typically immediately inside a missile skin is a layer of quiescent air and a lumped-capacitance model for individual elements along the surface. The heat transfer modelling accounts for material property changes of the skin and the convective heat flux as a function of vehicle orientation, flight speed and altitude, and location on the surface.

8.2.1.9 Cost Analysis

The missile cost is calculated for each new configuration. The equations are largely based on missile weight and material. The total cost is adjusted for inflation to the year 2024 and updated where required.

8.2.1.10 Optimisation Results

Typical design processes are laborious in the determination of the performance of candidate options, verifying that each discipline observes the constraints of other disciplines, overall constraints are met, and requirements are being addressed. The MDO framework automates much of the labour-intensive processes and allows the engineer to evaluate many candidate options consistent across all disciplines and adhere to all design constraints. This allows the designer to select the best vehicle as its performance is stacked against multiple goals. The resultant range versus cost pareto front is presented in Figure 8.12. Three designs were selected at chosen points along the pareto and their corresponding design parameters and performance metrics are shown in Table 8.2. To minimise cost while preserving stability, all three designs were optimised to have only three tail fins.

Concept #1: Located on the far right of the Pareto, represents the longest range, highest cost solution. It both maximises length and reduces nose cone fineness to maximise the size of the rocket motor.

Concept #2: Situated near a knee in the curve where the range performance drops rapidly as cost decreases. Size and weight being a major cost driver, this design is near the lower bound in length and relies on a high fineness nose to minimise drag

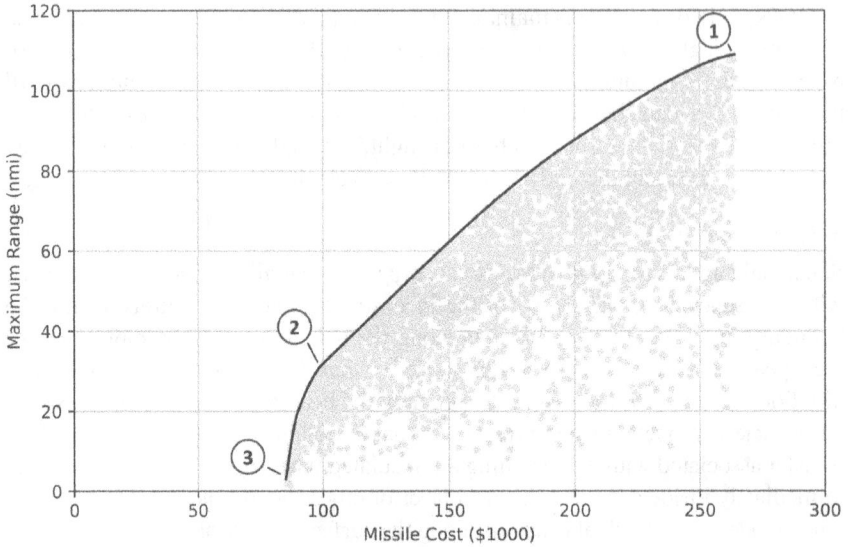

FIGURE 8.12   Range vs cost Pareto front.

TABLE 8.2
Potential Design Missile Parameters

	Concept 3	Concept 2	Concept 1
Launch angle, theta (deg)	61	64	42
Boost thrust (lbf)	180	763	831
Tail span (in)	15.9	11.4	12.8
Total length (in)	36.0	37.0	72.0
Tail deployment angle (deg)	139	90	94
Nose fineness	4.0	4.0	3.6
Canard span (in)	10.9	6.0	6.0
Canard chord fraction	0.6	0.5	0.5
Canard sweep (deg)	0.0	0.0	0.0
Number of tail fins	3.00	3.00	3.00
Total weight, launch (lb)	15.0	17.5	63.2
Cost ($)	$85,115	$100,126	$264,175
Gun launch exit mach	3.8	3.7	2.5
Max range	3	32	109
Time of flight (sec)	179	292	425
Average stability (Cal.)	0.1	0.2	-0.2

to maintain range performance. Any reduction in cost from this design results in an undersized rocket motor and a substantial decrease in performance.

Concept #3: Represents the lower end of the Pareto, where every effort has been employed to reduce cost. The rocket motor is just powerful enough to barely increase the missile speed once it is out of the gun, which results in an extremely short range.

8.3 MDO PROCESS

As budgets and schedules continue to be reduced, and performance desires continue to rise, it becomes critical to perform designs more rapidly and develop prototypes that are close to the final, tactical vehicle. The apparent contradiction requiring rapid development of a near optimal configuration can be addressed by conducting early design through a multidisciplinary lens. The approach of using an MDO process includes interconnected models for each discipline, quickly assessing the characteristics of candidates, and predicting their performance against the constraints and requirements. Modern computers allow for the rapid evaluation of thousands of candidate vehicles, and their results can be automatically examined against multiple requirements. The process starts with a wide design space using low-order tools for aerodynamic predictions, mass property approximations, cost estimates, etc. Subsequently, the results are inspected, and promising regions are further examined with higher fidelity estimates. The process continues until an extremely accurate result is generated for the nearly optimal, system-level design.

This example, and the entire chapter, has focused on the application of design thinking and the MDO process for flight vehicle modelling, design, and analyses. However, it is important to note that the application of the MDO practice to design thinking is applicable to virtually any design project. For any design problem, through some effort, all associated disciplines can be interconnected through a process/information flow to quantify performance against the goals and requirements of the user. Once automated, the technique allows for the rapid evaluation of a wide range of design options, which can be returned into the broader design thinking process to alter the process or home in on optimal designs. Although the application was for a weapon design, similar models could be generated for the land allocation of a farm, the design of a car transmission, or virtually any other engineering project.

NOTE

1 https://docs.h2o.ai/h2o/latest-stable/h2o-docs/data-science/deep-learning.html

REFERENCES

1. Plattner, H., Meinel, C., Leifer, L., "Design Thinking: Understand – Improve – Apply," Springer Heidelberg Dordrecht London, New York, 2011.
2. Schmit, L. A., "Structural Design by Systematic Synthesis," 2ndConference on Electronic Computation, American Society of Civil Engineers, New York, 1960, pp. 105–132.

3. Schmit, L. A., and Thornton, W. A., "Synthesis of an Airfoil at Supersonic Mach Number," NASA CR-144, Jan. 1965.

4. Schmit, L. A., "Structural Synthesis: Its Genesis and Development," AIAA Journal, Vol. 19, No. 10, 1981, pp. 1249–1263. doi: 10.2514/3.7859.

5. Schmit, L. A. Jr., "Structural Synthesis: Precursor and Catalyst, RecentExperiences in Multidisciplinary Analysis and Optimisation," NASACP-2337, 1984.

6. Haftka, R. T., "Automated Procedure for Design of Wing Structures to Satisfy Strength and Flutter Requirements," NASA Langley Research Center TN-D-7264, Hampton, VA, 1973.

7. Haftka, R. T., Starnes, J. H. Jr., Barton, F. W., and Dixon, S., "Comparison of Two Types of Optimisation Procedures for Flutter Requirements," AIAA Journal, Vol. 13, No. 10, 1975, pp. 1333–1339. doi: 10.2514/3.60545.

8. Haftka, R. T., "Optimisation of Flexible Wing Structures Subject to Strength and Induced Drag Constraints," AIAA Journal, Vol. 14, No. 8, 1977, pp. 1106–1977. doi: 10.2514/3.7400.

9. Haftka, R. T., and Shore, C. P., "Approximate Methods for Combined Thermal/Structural Design," NASA TP-1428, June 1979.

10. Ashley, H., "On Making Things the Best: Aeronautical Uses of Optimisation," Journal of Aircraft, Vol. 19, No. 1, 1982, pp. 5–28. doi: 10.2514/3.57350.

11. Green, J. A., "Aeroelastic Tailoring of Aft-Swept High-Aspect-Ratio Composite Wings," Journal of Aircraft, Vol. 24, No. 11, 1987, pp. 812–819. doi: 10.2514/3.45525.

12. Grossman, B., Gurdal, Z., Strauch, G. J., Eppard, W. M., and Haftka, R. T., "Integrated Aerodynamic/Structural Design of a Sailplane Wing," Journal of Aircraft, Vol. 25, No. 9, 1988, pp. 855–860. doi: 10.2514/3.45670.

13. Grossman, B., Haftka, R. T., Kao, P.-J., Polen, D. M., and Rais-Rohani, M., "Integrated Aerodynamic-Structural Design of a Transport Wing," Journal of Aircraft, Vol. 27, No. 12, 1990, pp. 1050–1056. doi: 10.2514/3.45980.

14. Livne, E., Schmit, L., and Friedmann, P., "Towards Integrated Multidisciplinary Synthesis of Actively Controlled Fiber Composite Wings," Journal of Aircraft, Vol. 27, No. 12, 1990, pp. 979–992.doi:10.2514/3.45972

15. Livne, E., "Integrated Aeroservoelastic Optimisation: Status and Direction," Journal of Aircraft, Vol. 36, No. 1, 1999, pp. 122–145. doi: 10.2514/2.2419.

16. Jansen, P., Perez, R. E., and Martins, J. R., "Aerostructural Optimisation of Nonplanar Lifting Surfaces," Journal of Aircraft, Vol. 47, No. 5, 2010, pp. 1491–1503. doi: 10.2514/1.44727.

17. Ning, S. A., and Kroo, I., "Multidisciplinary Considerations in theDesign of Wings and Wing Tip Devices," Journal of Aircraft, Vol. 47, No. 2, 2010, pp. 534–543. doi: 10.2514/1.41833.

18. Kroo, I. M., Altus, S., Braun, R. D., Gage, P. J., and Sobieski, I. P., "Multidisciplinary Optimisation Methods for Aircraft Preliminary Design," Proceedings of the AIAA 5th Symposium on Multidisciplinary Analysis and Optimisation, AIAA Paper 1994-4325, Panama City Beach, FL, 1994.

19. Manning, V. M., "Large-Scale Design of Supersonic Aircraft via Collaborative Optimisation," Ph.D. Thesis, Stanford University, Stanford, CA, 1999.

20. Antoine, N. E., and Kroo, I. M., "Framework for Aircraft Conceptual Design and Environmental Performance Studies," AIAA Journal, Vol. 43, No. 10, 2005, pp. 2100–2109. doi: 10.2514/1.13017.

21. Henderson, R. P., Martins, J. R. R. A., and Perez, R. E., "Aircraft Conceptual Design for Optimal Environmental Performance," The Aeronautical Journal, Vol. 116, No. 1175, Jan. 2012, pp. 1–22.

22. Alonso, J. J., and Colonno, M. R., "Multidisciplinary Optimisation with Applications to Sonic-Boom Minimization," Annual Review of Fluid Mechanics, Vol. 44, No. 1, 2012, pp. 505–526.

23. Martins, J. R. R., and Lambe, A., A. B., "Multidisciplinary Design Optimisation: A Survey of Architectures," AIAA Journal, Vol. 51, No. 9, September 2013, pp. 2049–2075.

24. McKay, M. D., Beckman, R. J., and Conover, W. J. "A Comparison of Three Methods for Selecting Values of Input Variables in the Analysis of Output from a Computer Code". Technometrics, American Statistical Association, 1979: 239–245.

9 Two Modes of Design Thinking Exhibited in Intuitive and Analytical Design Methods

Udo Kannengiesser and John S Gero

9.1 INTRODUCTION

Within design thinking there are two fundamental processes commonly understood as alternating between divergent and convergent phases. Divergent phases expand and explore the design space. Convergent phases focus on specific design issues and solutions (Design Council, 2024). Designers have been found to frequently shift between divergent and convergent phases (Goldschmidt, 2016). Based on the close alignment of the two phases with different cognitive thinking styles (Kochanowska et al., 2022), there has been strong interest in bringing models from cognitive science into design thinking (Hay et al., 2017; Howard et al., 2008). One of these models – the dual-system theory (Evans and Stanovich, 2013; Kahneman, 2011; Sloman, 1996; Stanovich and West, 2000) – has more recently been introduced in design thinking research (Gonçalves and Cash, 2021; Kannengiesser and Gero, 2019a; Lawrie et al., 2024; Moore et al., 2014). Dual-system theory posits that two systems or processes are involved in human cognition: system 1 for fast, intuitive thinking and system 2 for slow, analytic thinking. The two modes of thinking are viewed as interacting with each other. According to most accounts of dual-system theory, the principal role of system 2 thinking is to check if the results of system 1 thinking are suitable (Lawrie et al., 2024).

Empirical studies have provided quantitative evidence that designers frequently use system 1 thinking when generating design solutions (Kannengiesser and Gero, 2019b). This finding is consistent with general models of designing that emphasise the importance of intuitive thinking when problem and solution spaces are ill-defined (Pahl and Beitz, 2007). However, system 1 thinking is also seen as a source of bias for designers, as it can lead to fixation (i.e., premature commitment) on designs that do not perform well (Bulleit, 2013). There is a general view that such biases should be avoided or mitigated by design methods (Jimenez et al., 2024). Accordingly, methods are often categorised as either analytical or intuitive. Analytical methods aim to avoid system 1 thinking by structuring the process of generating design concepts as a sequence of deliberate steps. For example, morphological analysis (Zwicky, 1969) is a method that includes decomposing

 DOI: 10.1201/9781003487524-9

the functions of a design problem, producing alternative solutions for individual subfunctions, organising them in a morphological matrix, and combining them into an overall solution. Another analytical method is TRIZ (Altshuller, 2002) which provides a systematic way of producing design concepts using a set of 40 inventive principles and the resolution of contradictions. The application of this technique requires highly abstract, analytical thinking that is consistent with the notion of system 2 thinking.

In design thinking, intuitive design methods aim to make use of system 1 thinking for generating initial solutions, but then use system 2 thinking for analysing, evaluating, combining, or elaborating the concepts generated. This aims to mitigate potentially negative effects of biases associated with system 1 thinking. For example, brainstorming (Osborn 1953) occurs in two phases. In the first phase, system 1 thinking is used to quickly generate a large number of design solutions for a design problem, without caring about their feasibility or other qualities. In the second phase, the structures are analysed and associated with one another, which are activities predominantly based on system 2 thinking.

The extent to which these approaches can affect the two modes of design thinking has not been shown quantitatively. The goal of this chapter is to fill this gap in design thinking by measuring and statistically assessing the relative importance of system 1 and system 2 thinking when applying different concept generation techniques. The concept generation techniques studied are brainstorming, morphological analysis, and TRIZ.

9.2 DESIGN SESSIONS USING DIFFERENT CONCEPT GENERATION TECHNIQUES

Design thinking studies require a design task, means of capturing designers' thoughts, methods for analysing those thoughts, and methods for assessing the design result. Since thinking is not directly observable in the mind, we need to gain access to thinking in some secondary form. Humans articulate their thinking through their externalisations of thought. The most common form of thought externalisation is verbalisation – speaking, although there are others also. Studies of design thinking that use verbalisations are called *protocol studies*. Protocol studies capture designers' thoughts through their verbalisations. These verbalisations are transcribed and coded to turn the words into design-related categories. There are multiple coding schemes, each of which aims to provide some cognitive categorisation of the designer's thoughts.

Our analysis uses existing data (Kannengiesser et al., 2013) from protocol studies of 22 mechanical engineering students who were grouped into teams of two. Every team was given the same set of three design tasks, one per concept generation technique. All design tasks dealt with developing an assistive technology device and were created to be similar in concept, context, and complexity. In the first session, students were asked to use brainstorming to design a device to help disabled users open a stuck double-hung window without relying on electric power. In the second session, students were asked to use morphological analysis to design a device to help stroke patients, who are unable to perform bilateral tasks, with opening doors (Atman et al., 2008). In the third session, students were asked to use TRIZ to design a device to add to an existing hand/arm-powered wheelchair that will allow paraplegic wheelchair users to traverse a standard roadside curb unassisted.

During the design sessions, the students were asked to collaborate with their team members. They were instructed to intentionally and actively use one of the concept generation techniques, and to produce a design solution that meets the given design requirements within 45 minutes. All the design sessions were audio and video recorded for later analysis.

9.3   CODING AND ANALYSIS

The function-behaviour-structure (FBS) ontology (Gero, 1990; Gero and Kannengiesser, 2004, 2014) was used as the basis for coding the design protocols. It comprises six classes of design issues:

1. *Requirements* (R): These are the needs, demands, and constraints explicitly provided to the designer at the outset of a design process.
2. *Functions* (F): These are the purposes associated with the design object.
3. *Expected behaviours* (Be): These are attributes deemed necessary for the design object's expected interactions with the environment. They are used as criteria for assessing the performance of the design object.
4. *Behaviour derived from structure* (Bs) (or shorthand: structure behaviour): It comprises attributes measured or derived from observations of the design object and its interactions with the environment. They are compared with the expected behaviour (Be) for assessing the performance of the design object.
5. *Structure* (S): It includes the components of the design object and their relationships.
6. *Description* (D): It includes external representations of the design object produced during or at the end of the design process.

The designers' utterances captured in design protocols were segmented and coded as FBS design issues. Transformations of one design issue into another are represented by pairs of two consecutive FBS design issues. Every transformation represents a distinct instance of design thinking within the design process. Most of them are consistent with the eight processes defined within the FBS framework, shown in Figure 9.1, which are seen as fundamental in designing (Gero, 1990). They include:

1. *Formulation* (R → F and F → Be): establishes the problem space in terms of functions and expected behaviours, based on the requirements provided.

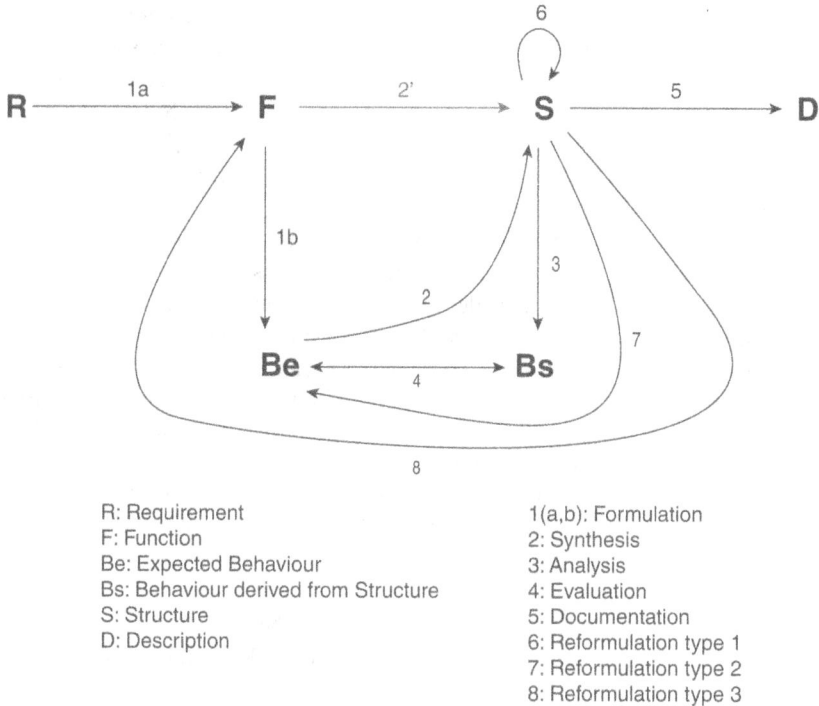

FIGURE 9.1 The FBS framework augmented with system 1 thinking (highlighted in blue colour). (After Kannengiesser and Gero, 2019a, 2019b).

2. *Synthesis* (Be → S): produces a structure that is to meet the expected behaviours.

3. *Analysis* (S → Bs): derives the "actual" behaviour from the synthesized structure.

4. *Evaluation* (Be ↔ Bs): assesses the design based on a comparison of "actual" and expected behaviours.

5. *Documentation* (S → D): produces a design description, mostly based on structure.

6. *Reformulation type 1* (S → S'): modifies the generated structure of the design.

7. *Reformulation type 2* (S → Be'): modifies the expected behaviours of the design.

8. *Reformulation type 3* (S → F'): modifies the functions of the design.

The eight processes in the FBS framework are viewed as instances of system 2 thinking. Recently, a ninth process has been added to account for system 1 thinking in design: the transformation of F into S (process 2' in blue in Figure 9.1) (Kannengiesser and Gero, 2019a). It is based on a reflex-like connection between a stimulus (F) and a response (S) learned by a designer through previous experiences, leading to increased cognitive efficiency when generating design solutions. It represents a "shortcut" for generating S directly from F, eliminating the system 2 thinking processes that would normally be needed to get to S: transforming F into Be (process 1b) and Be into S (process 2), then validating S by transforming it into Bs (process 3) followed by comparing Be and Bs (process 4). The F→S transformations are examples of system 1 thinking.

Design thinking may involve more processes than those shown in Figure 9.1, such as Be→D, F→F, and D→S. However, these processes represent either focus shifts that do not produce new design issues or transformations that are not viewed as fundamental to design thinking. Considering both fundamental and non-fundamental design thinking processes, we can analyse the occurrence of F→S transformations, i.e., system 1 thinking, relative to two different baselines (Kannengiesser and Gero, 2019b). The semantic baseline considers only those fundamental transformations of F defined in Figure 9.1 (i.e., F→S and F→Be). The syntactic baseline is a superset of the semantic baseline, considering all fundamental and non-fundamental transformations of F (whether or not they are defined in the FBS framework).

- Semantic baseline: Occurrence of any F→Y, where Y ∈ {Be, S}
- Syntactic baseline: Occurrence of any F→X, where X ∈ {R, F, Be, Bs, S, D}

In this model of design thinking, the number of F→S transformations relative to each baseline is a measure for the distributions of system 1 and system 2 thinking.

9.4 RESULTS

The percent occurrences of F→S for the different concept generation techniques are shown in Table 9.1. They are represented graphically in Figure 9.2.

Using a two-way ANOVA with repeated measures (chosen based on the dependent samples of student teams) and a significance level $\alpha = 5\%$, we found

TABLE 9.1

Percent Occurrences of F→S Relative to Syntactic and Semantic Baselines

	F→S Relative to the Syntactic Baseline (std dev) %	F→S Relative to the Semantic Baseline (std dev) %
Brainstorming	40.7 (15.8)	72.0 (15.8)
Morphological analysis	26.3 (11.5)	55.2 (15.6)
TRIZ	15.9 (9.3)	35.9 (14.5)

Note: Standard deviations are in brackets.

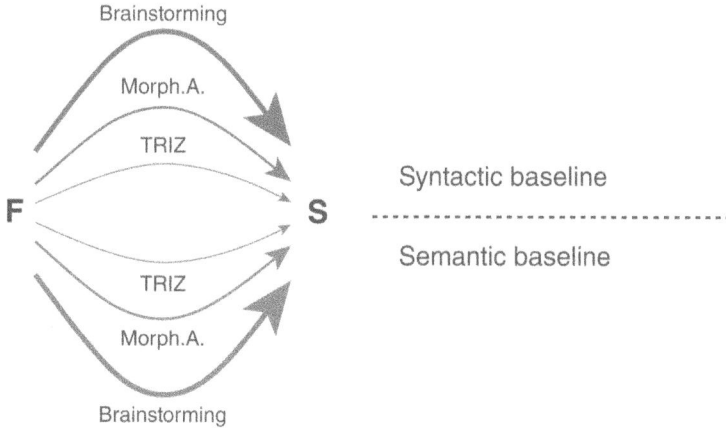

FIGURE 9.2 Graphical representation of the data in Table 9.1: The thickness of the arrows corresponds to the percent occurrences of F→S in the three concept generation techniques.

statistically significant differences across the three cohorts (brainstorming, morphological analysis and TRIZ), both regarding the syntactic (F = 5.49, p = 0.047) and the semantic baseline (F = 15.728, p = 0.004). Paired-samples t-tests were carried out to examine the differences between individual pairs of cohorts, as shown in Figures 9.3 (syntactic baseline) and 9.4 (semantic baseline). The results indicate that for the syntactic baseline there is a significant difference (α = 5%)

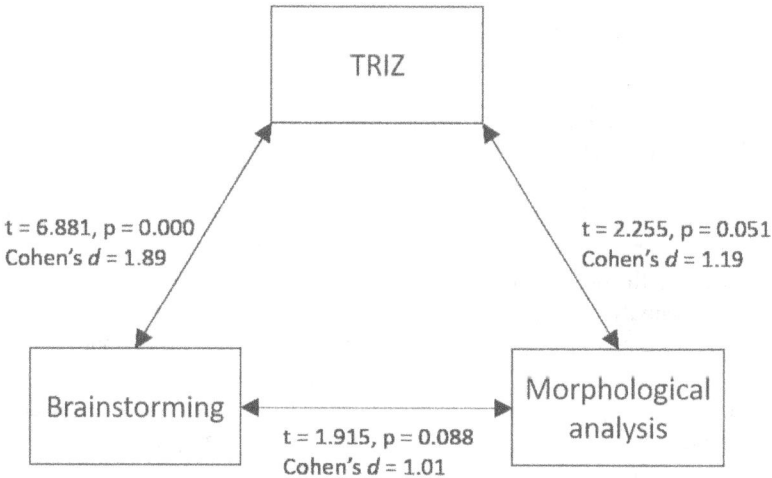

FIGURE 9.3 Results of paired-samples t-tests of F→S occurrences relative to the syntactic baseline, and associated effect sizes (calculated using Cohen's d).

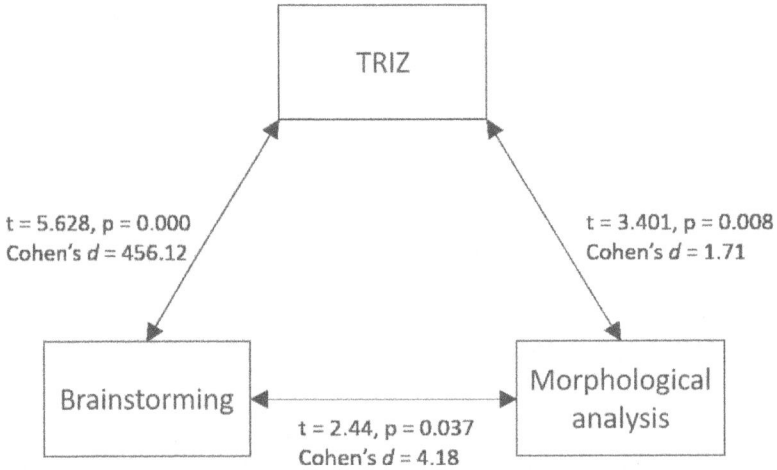

FIGURE 9.4 Results of paired-samples t-tests of F→S occurrences relative to the semantic baseline, and associated effect sizes (calculated using Cohen's d).

between brainstorming and TRIZ ($p = 0.000$). The difference between morphological analysis and TRIZ is close to being statistically significant ($p = 0.051$). For the semantic baseline, all three concept generation techniques are significantly different. The effect sizes for the paired t-tests, calculated using Cohen's d, are also shown in Figures 9.3 and 9.4. They are large for all pairs, and very large for brainstorming vs. TRIZ (semantic baseline).

9.5 CONCLUSION

The results of this research provide empirical evidence for two modes of design thinking – system 1 and system 2 thinking – in design sessions using intuitive and analytical design methods. The extent to which the two modes are used differs across the methods. Brainstorming (an intuitive design method) makes use of system 1 thinking most intensively, while TRIZ (an analytical design method) uses system 1 thinking much less. Morphological analysis (another analytical design method) sits in the middle. This fits with perceptions that the distinction between intuitive and analytical design methods is not clear-cut (Leenders et al., 2007). Design methods are on a spectrum characterised by different distributions of system 1 and system 2 thinking. In the case of morphological analysis, its position in the middle of that spectrum may be based on the ways in which partial structures are generated are not included in the method, which leaves room for system 1 thinking to get involved.

As the statistical results match common expectations about the connection between modes of design thinking and design methods, the relative occurrence of F→S in design protocols can be seen as a general metric that can be used to quantitatively analyse the distribution of system 1 and system 2 thinking in design methods. More design methods than the ones in this study may be examined using the same metric, independently of the domain of the design or the kinds of artefacts being designed. While the metric is useful for describing different modes of design thinking, it cannot explain their outcomes or their drivers. Current research is investigating neurological measures of the two modes of design thinking including metacognition, based on Electroencephalography (EEG) frequency band power (Vieira et al., 2023). This may help explain why designers switch between different modes of thinking.

ACKNOWLEDGEMENTS

This work has been supported in part by the US National Science Foundation, grant no. 1929896. Any opinions, findings, and conclusions or recommendations expressed in this material are those of the authors and do not necessarily reflect the views of the National Science Foundation.

REFERENCES

Altshuller G (2002) *40 Principles: TRIZ Keys to Technical Innovation*, Technical Innovation Center, Worcester.

Atman CJ, Kilgore D and McKenna AF (2008) Characterizing design learning through the use of language: A mixed-methods study of engineering designers. *Journal of Engineering Education* **97**(3): 309–326.

Bulleit WMB (2013) *Engineers Shouldn't Think Too Fast, Structures Congress 2013*Pittsburgh, PA..

Design Council (2024) *Framework for Innovation*, Design Council, UK. https://www.designcouncil.org.uk/our-resources/framework-for-innovation/ (last accessed 11 Aug 2024)

Evans JSBT and Stanovich KE (2013) Dual-process theories of higher cognition: Advancing the debate. *Perspectives on Psychological Science* **8**(3): 223–241.

Gero JS (1990) Design prototypes: A knowledge representation schema for design. *AI Magazine* **11**(4): 26–36.

Gero JS and Kannengiesser U (2004) The situated function-behaviour-structure framework. *Design Studies* **25**(4): 373–391.

Gero JS and Kannengiesser U (2014) The function-behaviour-structure ontology of design, *in* A Chakrabarti and LTM Blessing (eds) *An Anthology of Theories and Models of Design: Philosophy, Approaches and Empirical Explorations*, Springer, pp. 263–283.

Goldschmidt G (2016) Linkographic evidence for concurrent divergent and convergent thinking in creative design. *Creativity Research Journal* **28**(2): 115–122.

Gonçalves M and Cash P (2021) The life cycle of creative ideas: Towards a dual process theory of ideation. *Design Studies* **72**: 100988.

Hay L, Duffy AHB, McTeague C, Pidgeon LM, Vuletic T and Grealy M (2017) Towards a shared ontology: A generic classification of cognitive processes in conceptual design. *Design Science* **3**: e7.

Howard TJ, Culley SJ and Dekoninck E (2008) Describing the creative design process by the integration of engineering design and cognitive psychology literature. *Design Studies* **29**(2): 160–180.

Jimenez SH, Godot X, Petronijevic J, Lassagne M and Daille-Lefevre B (2024) Considering cognitive biases in design: An integrated approach. *Procedia Computer Science* **232**: 2800–2809.

Kahneman D (2011) *Thinking, Fast and Slow*, Penguin Books, London.

Kannengiesser U and Gero JS (2019a) Design thinking, fast and slow: A framework for Kahneman's dual-system theory in design. *Design Science* **5**: e10.

Kannengiesser U and Gero JS (2019b) Empirical evidence for Kahneman's system 1 and system 2 thinking in design, *in* Eriksson Y and Paetzold K (eds) *Human Behavior in Design*, UniBw München, Germany, pp. 89–100.

Kannengiesser U, Gero JS and Williams C (2013) What do the concept generation techniques of TRIZ, Morphological Analysis and Brainstorming have in common? *Proceedings of the 19th International Conference on Engineering Design (ICED13), Design for Harmonies, Vol. 7: Human Behaviour in Design*, Seoul, Korea.

Kochanowska M, Gagliardi WR and Ball J (2022) The double diamond model: In pursuit of simplicity and flexibility, *in* Raposo D, Neves J and Silva J (eds) *Perspectives on Design II*, Springer Series in Design and Innovation, vol. 16, Springer, Cham.

Lawrie E, Flus M, Olechowski A, Hay L and Wodehouse A (2024) From theory to practice: a roadmap for applying dual-process theory in design cognition research, *Journal of Engineering Design*, DOI: 10.1080/09544828.2024.2336837

Leenders RThAJ, van Engelen JML and Kratzer J (2007) Systematic design methods and the creative performance of new product teams: Do they contradict or complement each other? *Journal of Production Innovation Management* **24**(2): 166–179.

Moore D, Sauder J and Jin Y (2014) A dual-process analysis of design idea generation, *Proceedings of the ASME 2014 International Design Engineering Technical Conferences & Computers and Information in Engineering Conference IDETC/CIE 2014*, Buffalo, NY.

Osborn AF (1953) *Applied Imagination: Principles and Procedures of Creative Problem Solving*, Charles Scribner's Sons, New York.

Pahl G and Beitz W (2007) *Engineering Design: A Systematic Approach*, Springer, Berlin.

Sloman SA (1996) The empirical case for two systems of reasoning. *Psychological Bulletin* **119**(1): 3–22.

Stanovich KE and West RF (2000) Individual differences in reasoning: Implications for the rationality debate? *Behavioral and Brain Sciences* **23**(5): 645–665.

Vieira SLDS, Kannengiesser U and Gero JS (2023) Triple Process Theory and EEG Frequency Band Power: Metacognitive Brain Activity in Open Design, *Proceedings of the ASME 2023 International Design Engineering Technical Conferences and Computers and Information in Engineering Conference. Volume 6: 35th International Conference on Design Theory and Methodology (DTM)*, Boston, MA.

Zwicky F (1969) *Discovery, Invention, Research through the Morphological Approach*, Macmillan.

10 Co-Design

Meanings and Applications in Design and Engineering

Bahareh Shahri, Pierre Lubis, and Eujin Pei

10.1 INTRODUCTION

Co-Design and similar terms, such as stakeholder-driven design, design thinking with the users, collaborative design, collective design, participatory design, cooperative design, user innovation, and design for, with, and by users, are prevalent in contemporary design practice. These terms represent the evolving nature of the design process, where the involvement of multiple stakeholders, including end users, is critical to creating more effective and user-centred solutions. Among these, Co-Design is notably applied across a variety of disciplines to denote the contribution of varied roles and actors in the design process. The prefix 'co' is interpreted to reflect the concurrent, collaborative, cooperative, collective, or community-driven aspects of design processes (Ruiz & Maier, 2012), emphasising the importance of teamwork and shared responsibility in developing design solutions.

Whilst participation of people is a congenial quality of Co-Design, in computer science, electronics, and mechatronics engineering, Co-Design refers to the simultaneous design of hardware and software components in electronic systems, focusing on creating compatible and efficient systems (Ruiz & Maier, 2012) or other concurrent professional involvement, with no reference to end users or stakeholders.

Co-Design in design practices is a relatively new concept that is regarded both as a design process and a design methodology. As a design process, Co-Design emphasises the practical, hands-on aspects of involving and engaging various stakeholders in the problem-solving and solution offering processes. This pragmatic approach focuses on collaboration, iterative development, and real-world application to create designs that are more human-centred and effective. On the other hand, as a design methodology, Co-Design stems from a more philosophical domain, reflecting deeper theoretical considerations about the nature of design, participation, and collaboration. This perspective is concerned with the underlying principles and values that drive the collaborative design process, such as inclusivity, empowerment, and shared ownership of the outcomes. It explores how involving diverse voices and perspectives can lead to more innovative and equitable design solutions, fostering a sense of community and collective responsibility (Sanders & Stappers, 2014).

DOI: 10.1201/9781003487524-10

By combining these pragmatic and philosophical dimensions, Co-Design offers a comprehensive framework that not only aims to produce better design outcomes but also to transform the way design is perceived and practised. It encourages a shift from traditional, top-down design approaches to more democratic and participatory methods, where the voices of all stakeholders are valued and integrated into the design process (Büscher et al., 2009). This holistic approach makes Co-Design a powerful tool for addressing complex design challenges in various fields, from product development, marketing and technology to sport and coaching, medicine and public health, urban planning, community development and social innovation (Meroni et al. 2018; Björgvinsson et al., 2012).

Co-Design processes utilise a diverse array of tools depending on the discipline, project goals, and participant selection. These tools range from traditional methods including printed images, Post-it notes, and drawing instruments to more advanced technological devices such as AI, online canvases, and other digital platforms. This study explores recent innovations in tools that facilitate Co-Design sessions, tracking the increased reliance on technology and digital platforms.

In this chapter, we define Co-Design as a design process as well as a methodology and outline the usage of the term Co-Design across the engineering, technology and design disciplines. The aim is to provide a comprehensive overview of its applications. By examining the diverse contexts in which Co-Design processes and methodologies are employed, this study seeks to offer a range of practical applications and guidelines that can help practitioners and researchers address complex design problems. Through this exploration, the study hopes to foster a deeper understanding of Co-Design and its potential to drive innovation and improve the design process in various fields. In this pursuit, we benefited from a systematic review of literature from Scopus database and identified engineering research publications that align with Co-Design principles. The included literature was examined to understand discipline approaches of co-design, their scope of Co-Design processes, and the tools used in these studies. The next section will explore theoretical perspectives on the meanings and significance of Co-Design.

10.2   DESIGN THINKING AND CO-DESIGN

Co-Design is also referred as Design thinking with the users (de la Guía et al., 2017), which implies it is an extension of design thinking. Co-Design is compared to design thinking process (and sometimes with human-centred design) in various projects (Björgvinsson et al., 2012; Li et al., 2018). The most significant similarities and differences lay in the politics of design (Björgvinsson et al., 2012), enticing tacit knowledge (Spinuzzi, 2005) and empathy.

10.2.1   Design Politics and Power

Co-Design is a practice that challenges the traditional authority and established power dynamics of designers within the design process. This challenge is crucial, especially when we recognise that the design process itself is inherently political

(Buchanan, 2001; Low, 2019). Designers, as problem solvers, are expected to deliver appropriate and innovative solutions to various issues. Historically, this expectation has endowed designers with significant power to define, judge, and make decisions in the realm of design. In the collaborative setting of Co-Design, this dynamic shifts. Designers are no longer the sole decision-makers; instead, they must share their power with other stakeholders, including users, clients, and other members of the design team. This democratisation of the design process ensures that diverse perspectives are considered, leading to more inclusive and user-centred outcomes (Shapiro, 2005; Hirsch, 2016). The political nature of design means that decisions made during the process can reach resource-weak stakeholders (Bjögvinsson et al., 2012) and have widespread implications, influencing societal norms, behaviours, and even policy. By involving a broader range of voices, Co-Design seeks to address power imbalances and ensure that the resulting designs better reflect the needs and desires of all stakeholders, not just those of the designers (Beck, 2002).

Looking at the history of design, this is not the first time that designers have faced challenges to their authority and power. The entire concept of Design Thinking, in many ways, aimed to demystify a process that had been unknown for decades. Nigel Cross gives multiple examples of such tradition by quoting famous designers such as Richard Stevens, Jack Howe and Philippe Starck, believing in intuition in their design process and the black box thinking processes. Similarly, Bill Moggridge states his tacit approach to design. He says "Please forgive me if my theoretical explanations are limited; my excuse is that I am a designer. I can easily tell you what to do but find it difficult to articulate the rationale ..."(Moggridge & Atkinson, 2007). Designers have demonstrated essential skills such as modelling, sketching, visualisation, and communication but it is the intuition that has positioned them as unique problem solvers. Thus, in the common practice of design, design responsibilities have allowed designers to dominate the design discourse, often making critical decisions based on their expertise and intuition.

This positioning has often set them apart from other disciplines such as engineering and the sciences, which also claim problem-solving capabilities. Designers have for long relied on their intuition and expertise to navigate complex design problems. However, by the movements in formalising design processes, a desire to make design more systematic and accessible, the focus on intuition has received less attention. Pioneers in the field of design methodologies sought to break down the complexities of design into more manageable and teachable components. By doing so, they aimed to transform design from an esoteric art form, accessible only to a few with natural talent, into a disciplined practice that could be taught and learned systematically. Less than a century ago, design experts and educators such as Christopher Alexander, who is widely regarded as the father of design methodology, laid the foundation for systematic approaches to design (Jiang, 2019). They influence subsequent scholars and practitioners such as John Christopher Jones, design theorist and researcher with several publications on design methods and Bruce Archer who defined Design as a discipline. As the structured design processes grew, the reliance on intuition in design gradually diminished and was replaced by more evidence-based judgements and articulated and collective reasoning.

One of the key outcomes of such formalisations was the development of design thinking, a methodology and a process that provides a structured approach to problem-solving. Design thinking emphasises a deep understanding of user needs, ideating creative solutions, prototyping, and iterating based on feedback. This approach prioritises empathy, experimentation, and collaboration over the solitary intuition of an individual designer. The shift towards design thinking marked the beginning of democratising and demystifying design and design processes. By providing clear methodologies and tools, design thinking made the design process more transparent and replicable. However, it is worth mentioning that the nature of design problems as defined by the so-called second generation of design method-ologists in the 1970s (Miyoshi, 2021) differs from other disciplines. Design deals with 'ill-defined' or 'wicked' problems (Rittel & Webber, 1973), anticipating par-ticipatory approach and abductive reasoning (Miyoshi, 2021).

Design theorists have tried to address the wickedness of design problems and their complex dimensions, which brings back the earlier political discussion from a different angle. Matthews et al. aim to describe this quality of political dimen-sions of wicked problems. They highlight a socio-political reality that makes "resistance in accepting any new framing of the problem" in the eyes of the users and stakeholders.

> This has as much to do with the social, collaborative process of design – how designers facilitate participation with stakeholders – as it does the concepts that are generated as design propositions. In politically charged domains, any exter-nally imposed design recommendations, however brilliant they may be, are in as much danger of alienating stakeholders as they are of convincing them that change is desirable. And the marginalisation of any set of stakeholders who have partial ownership of the problem domain will only diminish any chance of a proposal's ultimate success.
>
> (Matthews et al., 2023)

Accordingly, co-design process is a negotiation-based search for settlements, which suspends rather than solves value conflicts. Thus, co-design may be con-strued as a form of diplomacy, which operates within certain political limits of designerly peace-making (Molnar & Palmås, 2022). Ideally, co-design facilitates mutual learning to create shared understanding and trust; developing communi-cation and participation through valuing unique abilities and interests (Rajapakse et al., 2021)

Revisiting tacit knowledge in the earlier discussion, Co-Design, emphasises the importance of recognising participants' tacit knowledge in the design process (Spinuzzi, 2005) as well as the designers'. Co-Design seeks not only the formal and explicit competencies of users but also their tacit knowledge (Latour, 2005 quoted in Bjögvinsson et al., 2012). Thus, the role of design and designers is to find ways to access and capture this tacit knowledge and skills as much as their own tacit and experiential knowledge. By doing so, their creativity is directed towards designing tools and techniques that elicit such knowledge.

Co-Design represents a significant change in design practice, shifting the traditional power dynamics and emphasising the political nature of the design process. Sanders and Stappers once mapped different methodologies against experts and participatory mindsets and participatory design sits as close as possible to the participatory and users' side (Figure 10.1). Similar to an earlier version, Sanders (2008)) assigned the horizontal dimension to the mindsets of those who practice and teach design research and the vertical dimension to the impetus of the design research approaches. Design-led groups develop research methods and tools from a design perspective; creative agendas, and the realm of possibilities, and research-led groups comprise design research methods and tools from a research perspective; scientific approach, empirical and evidence-based solutions. An expert mindset denotes treating the users as subjects to design for and a participatory mindset acknowledges the users as partners to co-create solutions with. We translate their participatory design and Human-centred Design to today Co-Design to Design thinking (which arguably shares similar values). Sanders and colleagues later (2014) indicated that the most recent term of co-creation/co-design, has replaced the previous term of participatory design. It is clear that "the terminology referring to participation also presents its own evolution" (de la Guia et al., 2017) while we observe how co-design is progressing from design thinking to more inclusion of design led outcome and users' active role in the design process.

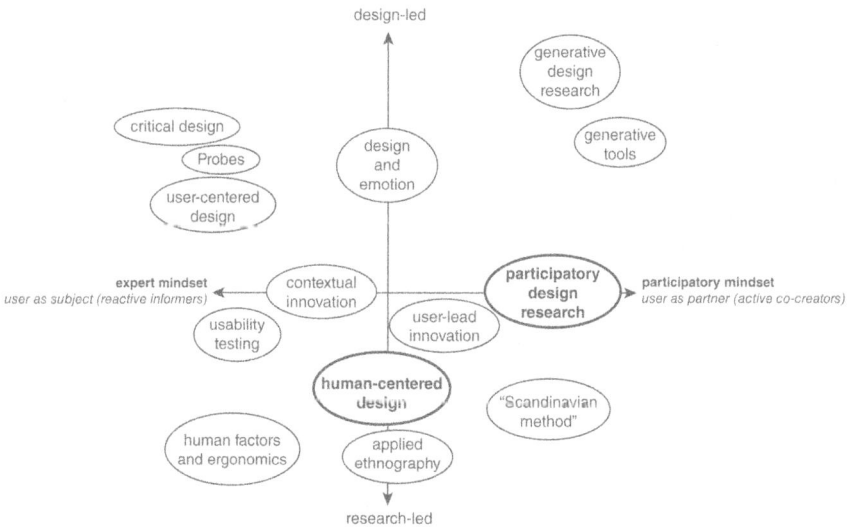

FIGURE 10.1 Positioning HCD and participatory design on the map of design research. (Prepared by the authors and adapted from Sanders and Stappers (2008)).

Following that trend, Co-Design presents itself at another level in the democratisation of Design. It encourages creative problem-solving and the generation of ideas that might not have emerged in a more hierarchical setting. It allows people from diverse backgrounds to engage in design activities, fostering a more inclusive environment where a variety of perspectives could contribute to the creative process. The political aspect of design becomes evident when considering how decisions made during the design process can affect various stakeholders. These decisions often reflect the values, priorities, and biases of the designers, which can have significant social, economic, and environmental implications. By involving multiple stakeholders, Co-Design seeks to democratise the design process, ensuring that a broader range of perspectives is considered. By challenging traditional power dynamics, Co-Design promotes a more participatory approach, leading to solutions that are not only innovative but also more attuned to the complexities of real-world and its actors.

10.3   TACIT KNOWLEDGE AND EMPATHY

As the focus of this writing to an extent, is the comparison of problem-solving processes in design and engineering, we can address the inherited differences of disciplinary values. One of the concerns in design which differentiates it from its counterpart engineering disciplines, is its focus on beyond functionality. Design has increasing presented concern for aesthetic appeal, style and expressive qualities (Postrel, 2003 quoted in Parsons, 2016), or as Christopher Alexander had put it the concern for appropriateness (Cross, 1982). Designers tackling ill-defined problems, intrinsically, cannot guarantee the success of their design solutions until it is being marketed and delivered to the end user. As such, they constantly improve design processes to minimise the risk of failures. To achieve this, they are obsessed with understanding users' needs and desires and look into ways to bring the user feedback closer to the front lines of the design process (Sanders & Stappers, 2014).

The empathy stage in Design Thinking and its tools are put forward to connect with the user at the initial stages. Similarly, in Co-Design, the empathic approach is at its core, but it is interwoven into all stages, from the beginning to the end. In this way, designers have the users and representatives of different stakeholders in the design process to make sure that beyond functionality aspects are constantly being checked and examined.

Empathic design (Leonard & Rayport, 1997) supports design teams in building creative understanding and involvement. Postma et al. (2012) explain that in empathic design approaches, the members of a design team (who may or may not be educated in design) adopt the role of people researchers and directly interact with users to ensure that their perspective is included in design. However, the inclusion of ideas and knowledge does not guarantee the flow of an empathic approach in other stages. Recent ideas such as "User-generated content" (Plank, 2016) highlight the importance of the correct interpretation of user contributions (de la Guia et al, 2017).

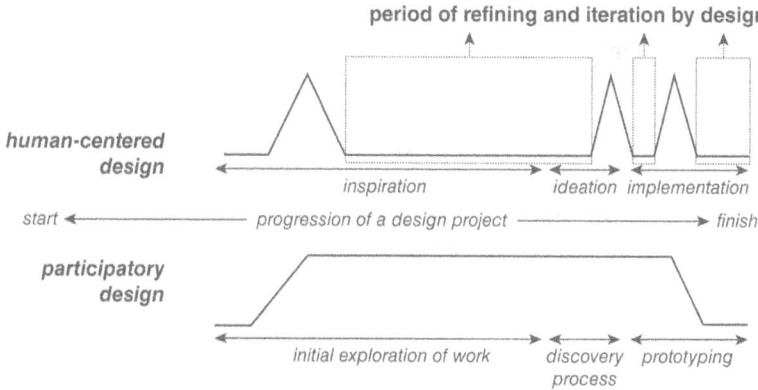

FIGURE 10.2 Comparing HCD to PD in the progression of a design project. (Prepared by the authors and based on Spinuzzi (2005) and IDEO.org (2015)).

Another comparison between Design thinking and Co-Design clarifies the role of designers in eliciting knowledge, including tacit knowledge, at the stages of the design process. Lubis et al. (2024) compared the contact time between designers and participants in participatory design and in human-centred design (HCD), as observed in Figure 10.2. The vertical axis indicates the level of contact time between designers and participants. The study highlighted the differences in the period and intervals of collaboration. In HCD, there is a shorter contact time with the participants, whereas in the Participatory model design stages all happen together, and the model does not allow HCD's long periods of refining and iteration by the designers. Hence, in participatory processes, the constant engagement with the stakeholders provides a great amount of data to process and turn into information and knowledge.

10.4   CO-DESIGN AS A METHODOLOGY

Methodology is defined around how to go about finding out knowledge and answers to our questions, and as Gray and Malins explained, it points to philosophical questions of how we see the world (Ontology) and the nature of reality and knowable, and how human-being/researcher can find out about the reality (Gray & Malins, 2004). Referring to the trilogy of Ontology, Epistemology and Methodology by Gray and Malins (2004), we can define Co-Design as a methodology which is based on an ontology formed by the belief in multi-realities. As such, finding reality can only be achieved collectively, a paradigm which stands for the position that the inquirer would be unsuccessful in its findings individually and the relationship between the inquirer and the knowable is complex, conflicting, and situated (Spinuzzi, 2005). In the constructivist epistemology and more open to interpretation views, participants' knowledge is "valorised rather than deprecated" (Spinuzzi, 2005), and their perspectives is shaped "in a dynamic system of people, practices, artifacts, communities, and institutional practices"

(Mirel, 1998, p. 13 quoted by Spinuzzi, 2005). Knowing in this context will be possible through reflection, allowing for voicing what cannot be articulated.

From a methodological standpoint, Co-Design reflects deeper theoretical considerations about the nature of design, participation, and collaboration. Co-Design methodology is rooted in the principle of inclusivity. It seeks to democratise the design process by ensuring that all voices, especially those of marginalised or less-heard groups, are represented and valued. This inclusive approach can lead to more equitable and just design outcomes. People have their voices including varied voices. Therefore, treating "users as resources", (Manzini, 2015 quoted by Meroni et al., 2018) as well as valuing the "wisdom of crowds" or collective judgment, which goes back to the writings of Aristotle. This entails accepting the premise that the users themselves are more knowledgeable than designers in their own experience and the context of usage.

Moreover, Co-Design believes that users are creative in contributing to problem solving. By involving stakeholders in the design process, Co-Design empowers individuals and communities to take an active role in shaping the solutions that affect their lives (Meroni et al., 2018). This empowerment can foster a sense of confidence, ownership and agency, enhancing the relevance and sustainability of the design outcomes (Jagtap, 2022a). This collective responsibility can lead to stronger commitment and support from all stakeholders, facilitating smoother implementation and greater long-term success. In other words, users become partners, as experts, to share their experience.

Co-Design methodology involves reflecting on the fundamental principles and values that guide the collaborative design process (Spinuzzi, 2005). This reflection helps to ensure that the design practice aligns with broader societal goals, such as social justice, environmental sustainability, and ethical responsibility. As such, it opens new arena for designers to tackle problems in their world that they previously could not take or imagine as their responsibilities. Thus, this is an epistemological difference between what designers knew about their role with the world. St John and Akama (2022) suggest co-design as an alternative *way of being* a design researcher. By shifting away from de-personalised accounts of research that emphasises roles, skills, processes, and methodologies, co-design is rather a co-ontological way of becoming, which troubles research traditions of replicability and generalisability.

Another ontological aspect of Co-Design is its connection with decolonial movements and the revitalisation of local and indigenous knowledge and practices. As disciplines shift away from their Euro-centric traditions and broaden to include local and regional knowledge, the search for appropriate and empathetic solutions becomes increasingly complex. Transitioning from an established Co-Design approach in Scandinavian countries to a decolonised practice, Co-Design focuses on addressing systematic inequities and resolving them at social and political levels, an area in the interest of social innovation. As Ehn thoroughly noted while theorising early Co-Design platforms, there has been "a shift from participatory design aimed at working in companies to a participatory design devoted to enhancing processes of empowerment within communities"

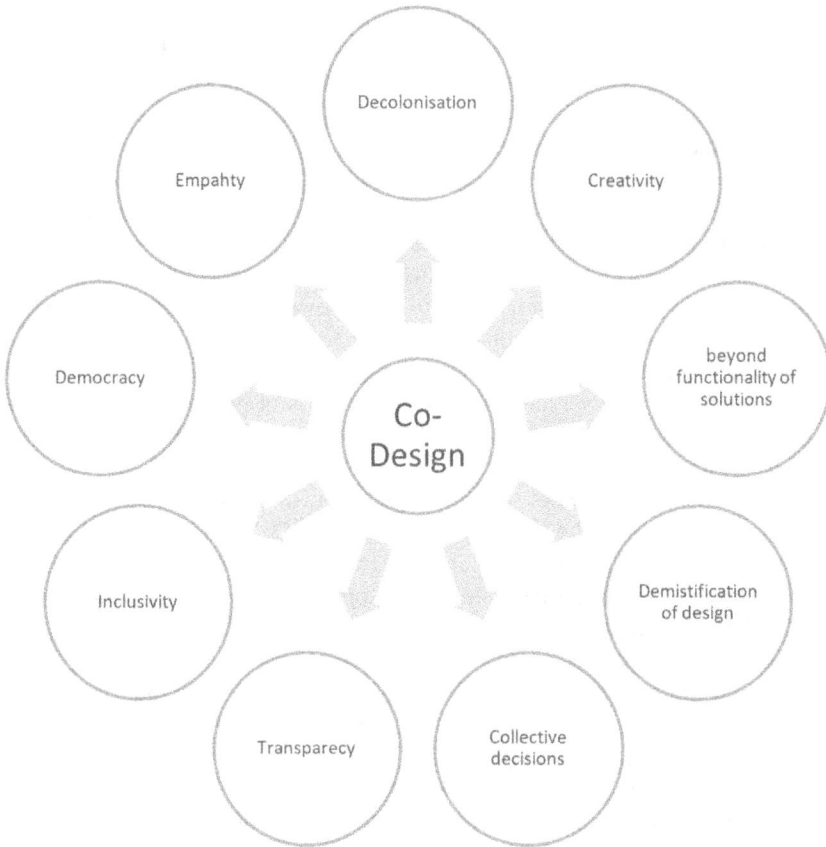

FIGURE 10.3 Values associated with Co-Design process.

(Ehn, 2008)(Ehn, 2008 quoted in Meroni et al., 2018). This has led to enabling democratised innovation today (Von Hippel, 2005, Björgvinsson et al., 2010). In coming reviews of literature, we investigate how co-design in engineering and technology reaches to this point and tackles inequities.

In summary, the discussed values of Co-Design can be visualised as Figure 10.3.

10.5 CO-DESIGN AS A DESIGN PROCESS

Common practice of involvement of the participants and subject matters in design processes has been parashooting in and out (Ranabahu et al., 2024) when the secondary information is not enough and firsthand information and insight is required (Voss, et al., 2009) whether it is at the "fuzzy front-end" or usabilty studies and testing. Co-Design tries to change that with recognising the value of the stakeholder's input in all stages of the process.

Manzini (2016, p. 58) defines Co-design as a complex, contradictory, sometimes antagonistic process. When design opens to a diverse range of people, it is clear that opinions, ideas, and desires would be different and sometimes conflicting and generate tensions (Molnar & Palmås, 2022). Moreover, from the perspective of a political theory realist, co-design can be understood as a zero-sum game, with winners and losers (Von Busch & Palmås, 2023). Therefore, it is the job of the designer to facilitate the discussion for maximum benefits. In Co-Design processes, the time of the designing is distributed between the stakeholders and designers, but the stages follow the mainstream ones. For example, Meroni et al. (2018), in their book *Massive codesign: A proposal for a collaborative design framework,* elaborate on the collaborative aspects of each stage of the design process following the Double Diamond Design Process by the Design Council. They believe in both divergent and convergent parts of process; staging out the processes of solving problems and offering innovation, and collaboration supports better understanding of the problem, gaining insights, expansion of creativity and realistic assessment of solution collectively. In other words, collaboration should be interwoven in the four D parts of Discover, Define, Develop and Deliver (Figure 10.4). At each stage, users and designers contribute to problem-solving, and they move from one side, with more "topic-driven" activities that refer to problem/situation defining to the other side which is more "concept-driven"; the problem-solving activities. Following this adaption, Mernoi et al. (2018) introduced and examined the Collaborative Design Framework with four dimensions of guidance roles for designers and involvement of the users in co-design processes (Figure 10.5). The Y axis added to the Double Diamond differentiates between steering and facilitating roles of designers. In Co-Design process steering is more encouraging on idealistic scenarios, generating and envisioning and

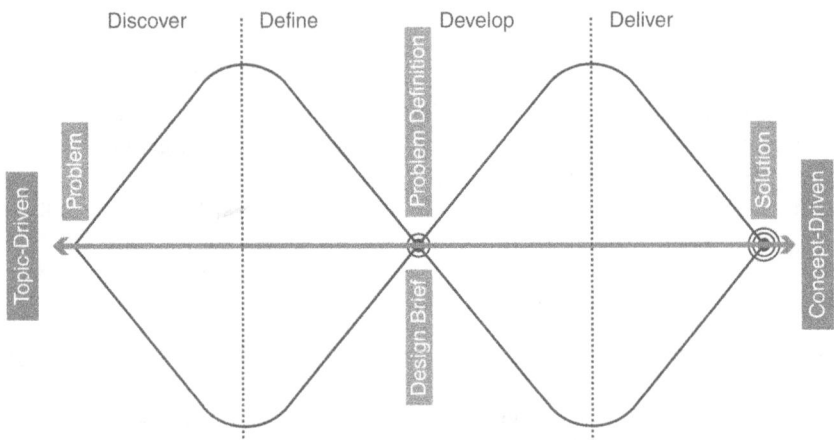

FIGURE 10.4 The double diamond scheme elaborated with two polarities about the subject matter of design by Meroni et al. (2018).

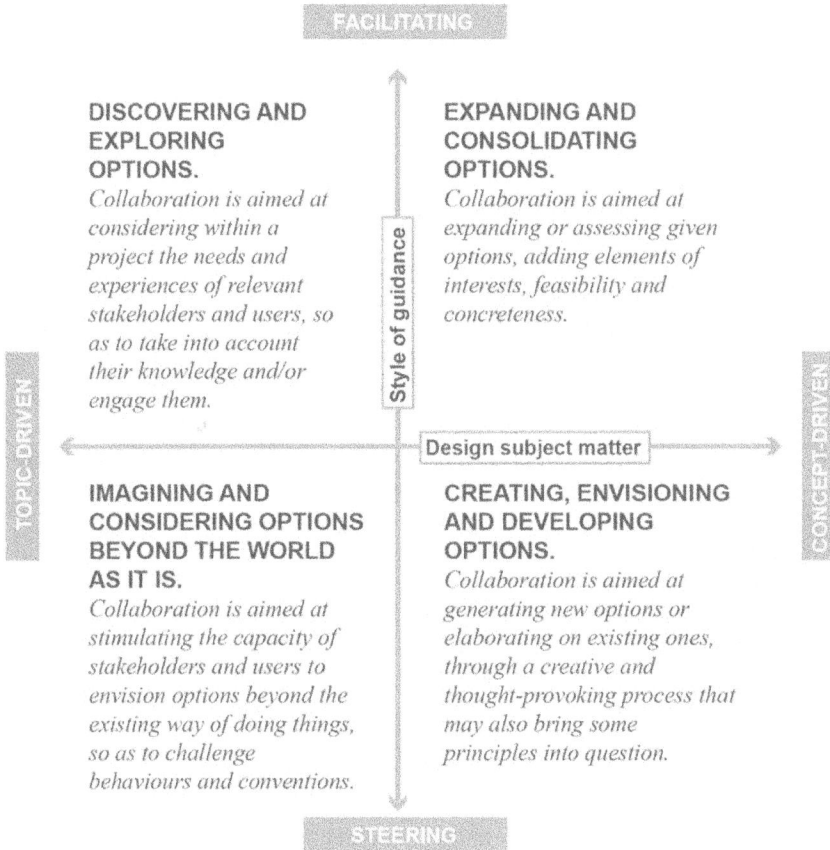

FIGURE 10.5   The collaborative design framework by Meroni et al. (2018).

facilitating in their guidance style. In our research, the format of co-design processes will be searched later among the literature.

Meroni et al. emphasised the role of aiding objects/systems in the process. They borrowed the term Boundary Object from sociology (Star, 2010) to explain the difference between objects/systems used in co-design process. They used two words of "prototypes" and "tools" to differentiate the purpose of materials deployed in co-design stages. Prototypes represent the embodied output of ideas, and tools include any apparatus used in all stages to connect designers and participants; aiming to help define the situations, imagine possibilities, bring empathy and in general synchronise Designers' and Participants' thinking (Meroni et al., 2018). In other words, the same objects and materials might be used as prototypes and tools, but their role and purpose are distinct. In this study, we focus on tools as they are well connected with the types of participants and dynamics of their interactions in co-design processes. The design and deployment of tools are determined by the co-design approach and knowing "who the relevant actors are, what their interests and commitments are, and how they relate to each other" as well as the format of questions, answers and literary genres used (Voss, et al., 2009).

10.6   CO-DESIGN SYSTEMATIC REVIEW PROCESS

As this study focuses on the meaning and application of Co-Design, we aimed to examine its usage in closely design-related disciplines with a problem-solving agenda, such as engineering and technology. To do so, after undertaking a literature review on Co-Design principles in design practices, we expanded the study to include a systematic literature review via Scopus database. Scopus as a scholarly database across a wide variety of engineering and technology disciplines enabled us a search among various subjects and disciplines.

Using similar techniques applied by Ruiz et al. (2012), Bazzano et al. (2017), de la Guía et al. (2017), and Jagtap (2022b), we started searching with the main keyword of Co-Design and its different formats. The review protocol (Figure 10.6) was an adapted version of PRISMA-P framework (Moher et al., 2009).

Codesign, co-design, participatory design and collaborative were the keywords used to identify the appropriate literature in the database. The search carried out within the article titles, abstracts, and keywords and this resulted in over 36 thousand hits. Through various filtering, such as subject areas (limited

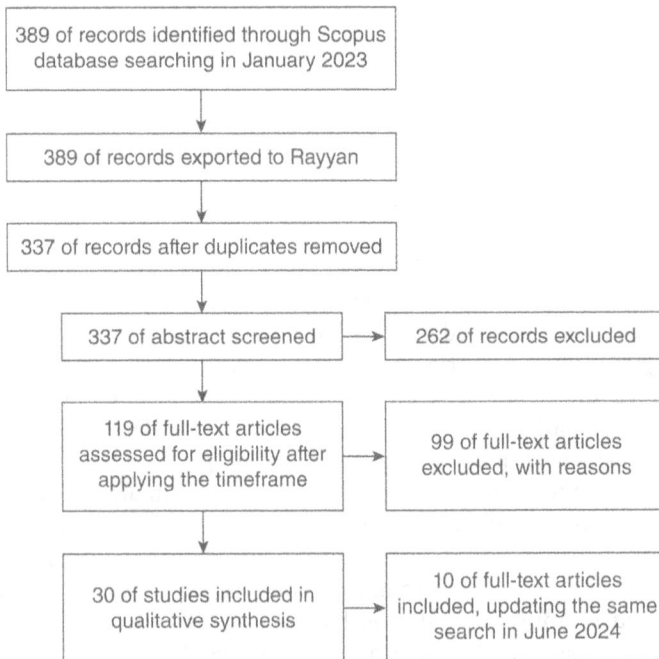

FIGURE 10.6   Protocol of systematic review adapted version of PRIDMA-P.

to engineering field), document type (limited to conference papers and articles), publication types (limited to final), keyword (excluded a number of pure engineering terms such as, hardware/software co-design, HW/SW co-design, embedded systems, computer hardware, computer software, etc), language (limited to English), we were able to reduce the results to 389 items on January 11th, 2023, and 464 on June 17th, 2024. The record of this advanced search can be found in the Appendix 1. Records were exported into Rayyan, a web-based collaborative systematic review platform (Ouzzani et al., 2016) and checked for any duplicates. Two reviewers independently assessed the titles, abstracts, and full texts against the criteria of inclusion. All disagreements were discussed between the two reviewers to reach the final decision. The criteria of inclusion and exclusion were (1) usage of Co-Design full process from the initial stages to the end; (2) present a new version of the co-design model; (3) involve a range of stakeholders including the end users, or (3) specifically address the tools and techniques in the study. As such, the list comprised 127 records, 119 in January 2023 plus 10 in June 2024. To reach a manageable number of articles we limited the search to over 2020, after the pandemic hit with the aim of identifying more technology-aided processes. Our justification was that the pandemic in 2020 induced new platforms of collaborations and Co-Design had to steer away from its traditional methods. We shortlisted the Scopus search result into a sizable amount to conduct the systematic review. Therefore, from a total of 129 articles, 30 were included in the final review and limited to articles published between 2020 and June 2024.

10.7 CO-DESIGN AND DISCIPLINES

Furthermore, the analysis of our Scopus review demonstrates the escalation of publications with the Co-Design approach in recent years (Figure 10.7). Our reviews of literature show that the term Co-Design is increasingly being used in a manner akin to its application in design disciplines, especially in pedagogical research publications by engineers (Aboutajedyne et al., 2022; Dong et al., 2022; Smith et al., 2022), as well as in fields such as environmental and urban design (Liu, 2022) and usability studies of technology (Kwieciński & Słyk, 2023). This trend highlights the growing recognition of the value of collaboration and multi-disciplinarity.

Initial search results presented a wide range of distribution of Co-Design publications across 25 disciplines (Figure 10.8), with the highest record in engineering,

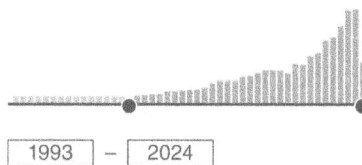

FIGURE 10.7   The increasing number of publications in Co-Design from 1993 to mid-2024 (Scoupus search).

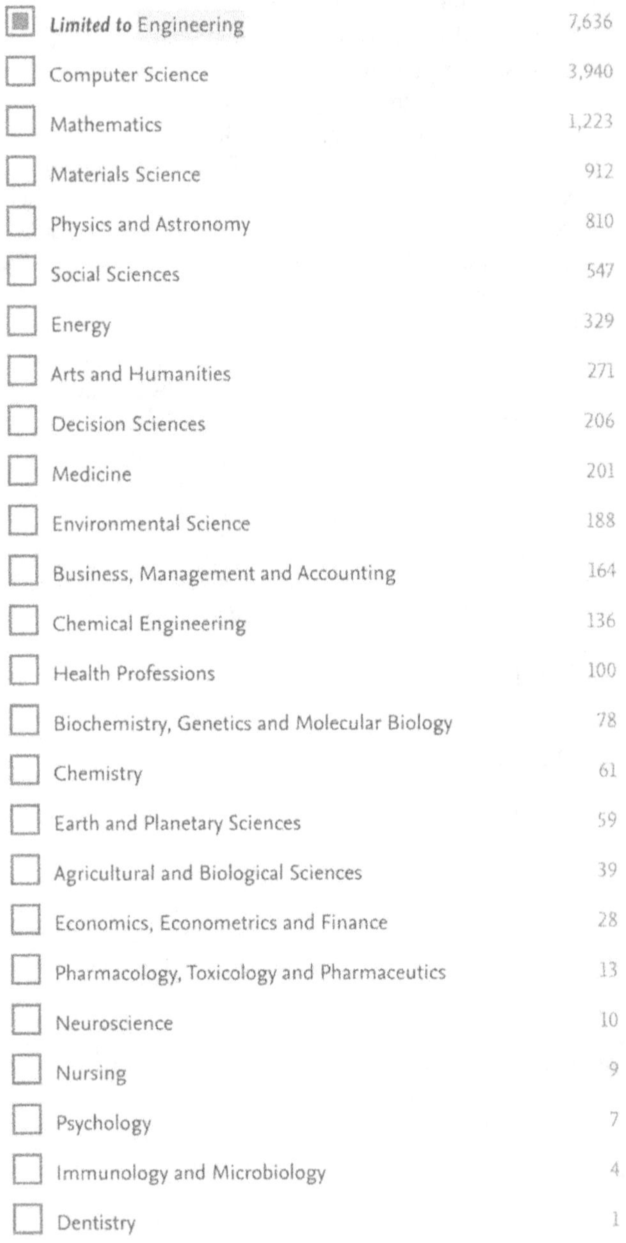

☒	*Limited to* Engineering	7,636
☐	Computer Science	3,940
☐	Mathematics	1,223
☐	Materials Science	912
☐	Physics and Astronomy	810
☐	Social Sciences	547
☐	Energy	329
☐	Arts and Humanities	271
☐	Decision Sciences	206
☐	Medicine	201
☐	Environmental Science	188
☐	Business, Management and Accounting	164
☐	Chemical Engineering	136
☐	Health Professions	100
☐	Biochemistry, Genetics and Molecular Biology	78
☐	Chemistry	61
☐	Earth and Planetary Sciences	59
☐	Agricultural and Biological Sciences	39
☐	Economics, Econometrics and Finance	28
☐	Pharmacology, Toxicology and Pharmaceutics	13
☐	Neuroscience	10
☐	Nursing	9
☐	Psychology	7
☐	Immunology and Microbiology	4
☐	Dentistry	1

FIGURE 10.8 Scopus records of appearing Co-Design and participatory terms in different subject of studies including engineering.

nearly double times more than records in the next discipline, Computer science. In engineering, the first shortlist presents topics in the Built environment including Architecture and Urban design (#24), Mechatronics Engineering (#1), Industrial Engineering (#4), Engineering Design (#2), Robotics (#1), Augmented Reality (#1) and Machine Learning (#1)- some of the papers coded for more than one discipline.

This overview gives us the full picture of how Co-Design is expanding in contemporary academic research. To identify significant approaches and principles, the full text of 30 articles (Appendix 2) were reviewed and the following themes were then extracted: (1) participant characteristics and their interactions, (2) co-design process, (3) co-design tools, and (4) politics of politics. These themes will be explained with the reference to the relevant articles subsequently.

10.8   PARTICIPANTS CHARACTERISTICS AND THEIR INTERACTIONS

In the selected studies, participants represented the general public, public sector, NGOs, local residentials, low-income communities, disadvantaged social groups, disabled communities, caregivers, refugees, long-term care residents, team workers, and professional experts. Co-Design processes presented in our search was not limited to certain nationalities or borders (Francis & Murtha, 2023; Miranda & McClam, 2023). Studies were taken place in South Africa (Govender & Loggia, 2023), Jordan (Albadra et al., 2021), Italy (Collina et al., 2022), India (Sharmin & Khalid, 2022), Nepal (Ramstad et al., 2020)), and Ecuador (Enrique et al., 2016). The potential of designing with and for nonhumans to reinvigorate modes of co-living and support existing habitats has also been investigated (Saeidi et al., 2023).

One of the conclusions raised in Collina et al. (2022) is that engaging with the end users, namely the managers, during the co-design process, despite its advantages, was challenging. Participants stated distinct needs, frequently incompatible and conflicting, and the researchers highlighted "the importance of establishing shared stakeholder conceptions and developing tools that would enable non-designers to participate in the project's debate" (Collina et al., 2022). The complexity of designing Co-Design is to overcome the maximum participation of the attendees, elicit tacit knowledge and maximise communication and understanding between the stakeholders at every stage of the process (Ben Guefrech et al., 2023). This includes how participants feel comfortable about sharing their ideas and opinions. As such, design of the process and tools have strongly influenced the outcome. Tools and methods should be matched with the experience and interest of the participants as well as their feelings during the participation, feeling of shyness, hesitation, or passion.

10.9   CO-DESIGN PROCESSES

Co-Design processes, similar to other design processes, benefit from creativity and adaptations to make them more appropriate and optimal (Wilson et al., 2015; Govender & Loggia, 2023; Madden et al., 2014; Hendriks et al., 2015;

Flatscher & Riel, 2016). In our review, a wide range of stages from sharing the brief, gaining insights, ideation, to realisation, feasibility study and user testing was spotted.

In Sharmin and Kahlid's housing project in India, their community participation has run throughout the project and included occupants in "decision-making, implementation, benefits acquisition, and evaluation" (Sharmin & Khalid, 2022). The architect team's decisions were checked and communicated with the community over and over. Through four phases of Project Initiation, Housing Design, Project Implementation and Evaluation and Redesign, participation of the community valued and cultural and social factors were considered, including the agreement of the community on vulnerability of some members than others, how the built can be resourced locally, the labour sourced locally to bring more income to the community and constant feedback from the early occupants for coming iterations and improvements.

The Co-Design processes with divergence and convergent phases were a recurring theme in the studies as it appeared in the work of Liu et al. (2023). Borrowing the Meinel et al.'s model (Meinel et al., 2011), they reached for users Points of View (PoV) (Figure 10.9). These characteristics in Co-Design process were important in their approach and considered as "a general participatory design methodology", staging out "interviewing, generating Points of View (PoV) for iterations and final outputs" (Lui et al., 2023).

In a novel practice, in a design engineering process, Rodrigues and colleagues combined Co-Design with the House of Quality Matrix to promote users' need, social inclusion, and multi-disciplinarity in the design of assistive devices (Rodrigues et al., 2023).

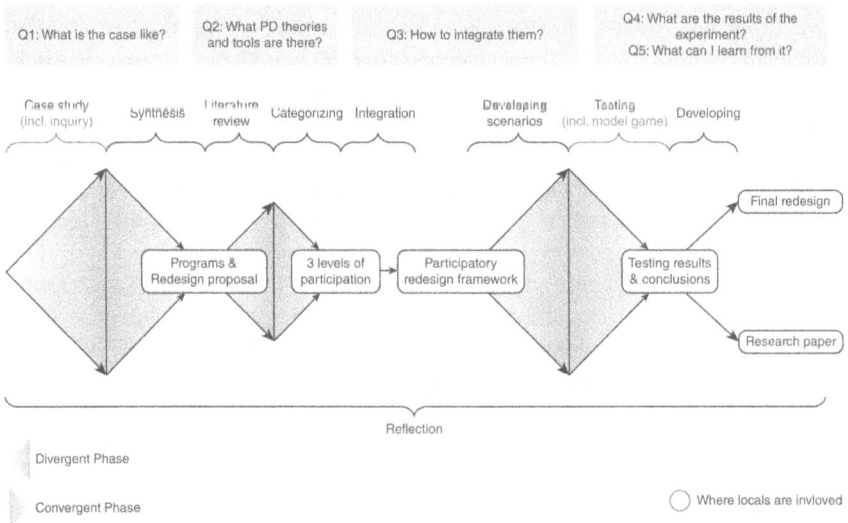

FIGURE 10.9   Presentation of divergent and convergent phases in Liu et al. study (2023).

In robotics, Co-Design has been used to engage with the end users, borrowing IDEO's basic HCD model, Hearing, Creating and Delivering, referencing Cross (2011). The researchers believe this approach, engaged them with the user through a participatory design process, which is supported by activities such as building listening skills, running workshops, and implementing ideas (Casiddu et al., 2022).

10.9.1   CO-DESIGN TOOLS

While Co-Design has been proven beneficial to achieve more affordable and efficient solutions (Brysch et al., 2023), it has benefitted from various digital tools which helped lower the requirements of time, language skill, and learning capacity for participants, and in turn made Co-Design processes to be easily repeatable contexts (Liu et al., 2023).

Manual tools	Pen and pencil, papers and drawing boards, sort cards, print images, sketching tools, maps, diagrams, storyboards, toys, acting accessories, cardboards, 3D models and prototypes tools
Digital tools	Screen digital devices and tablets, big screens, shared screens, VR and AR headsets, software programmes, shared databased AI

Tools and techniques in Co-Design processes play a significant role in enticing conversations and interactions between the participants and designers (Skoury et al., 2024). Tools function as icebreakers, triggers of conversation, prompts, facilitators of communication, brainstorming and ideation, visualisation of ideas, and recording agents. Some of these are adopted from other disciplines, such as psychology and ethnography, and adapted, and some are developed and introduced by designers themselves. For example, cultural probes (Gaver et al., 1999) introduced by Gaver and his colleagues at the Royal College of Arts was a package with maps, postcards, cameras and stationaries mainly designed to provoke novel thinking, inspirational responses from participants with no design background. The package allowed easy systems for recording and engagement when designers were not present.

In our research, the cultural probe was identified in the work of Riberio and Roque (2022), which introduced play and gamified aspects to Co-design in their maintenance solution Co-Design workshop on an aircraft business. They state "playful probing espouses similar principles of cultural probes while exploiting games as a research tool to enable learning and data collection. However, we still must identify how to research risk scenarios. The answer can be a simulation game as an enabler of participatory context. However, we do not know how we can create such a participatory context using a simulation game".

In architectural engineering projects (Collina et al., 2022), the tools consisted of maps, plans, transparent graphic boards, sheets and marking and colouring devices, and catalogues of possible and available choices of designs, such as on furniture, colours and technologies to be used in the space. Techniques and methods of engagements consisted of individual reflection time, group work, debate and discussion.

Digital tools and new technologies as solutions to foster collaboration were identified in the literature. Ben Guefrech et al. (2023) classified the mediating tools based the technology being used in the session. The tools can be tangible artefacts (all the physical objects presented in the room such as papers, pens, pencils, physical models, sketches, post-it notes, etc), digital artefacts (any kind of media displayed on a surface such as a TV screen, laptop, tablet, smartphone or any ICT tool. The later artefacts can be used for the presentation of text, pictures or images of the product, with the most frequent case in their study. In addition, videos and mixed artefacts, including mediums such as AR and digital artefacts, are projected on real space and objects.

Albadra et al. benefited in their study from a similar range plus prototyping equipment. The tools included pens and papers, a deck of cards, and plasticine at the earlier stage, and continued with 2D and 3D architectural drawings produced using computer models and physical prototypes with laser cutting machines, 3D printers, and eventually using VR equipment for final feedback (Albadra et al., 2021). Using these technical tools were possible by having relevant skills in the designers' team.

While both mechanical and computer-aided tools were found to be beneficial, in Ehab et al. project, participants reported a preference for the direct implementation of VR in BIM software. This research highlights the potential of VR technologies—specifically, BIM and game engines—as a co-design approach for public and social spaces in urban environments. It also identifies limitations and future research opportunities in adopting these methods for participatory design (Ehab et al., 2023).

At the same time, the emerge of apps and the generative tools in facilitating collaborations can be observed. To solve the top-down procedure of decision making in architectural design, Lui introduced an app with an interface that facilitate the complication of design decisions for non-designers and housing residences (Liu, 2022). In the same manner, a generative tool used in architectural project only at ideating session and testing the desired designs, layout of the house. (Kwieciński & Słyk, 2023). Both tools allow non-face to face collaboration that presents a future direction in collaborative design.

AI seems to be an emerging tool to use in Co-Design. In larger-scale public architecture projects, the client is frequently represented by a community that embraces numerous stakeholders. The scale, social diversity, and political layers of such collective clients make their interaction with architects challenging. A solution to address this challenge is using new information technologies that automate design interactions on an urban scale through crowdsourcing and artificial intelligence (AI) technologies. However, since such technologies have not yet been applied and tested in field conditions, it remains unknown how communities interact with such systems and whether useful concept designs can be produced in AI. To fill this gap in the literature, this paper reports the results of a case study architecture project where a novel crowdsourcing system was used to automate interactions with a community (Dortheimer et al., 2023).

10.9.2 POLITICS OF POLITICS

While Co-Design in most articles reinforces democratic (Corrado et al., 2024), sustainable (Liu et al., 2023; De Medici et al., 2023) and socio-cultural (Albadra et al., 2021) dimensions to making decisions, only two papers discussed the political aspects explicitly and in detail. Wickerson reminds us that in engineering and its education, people and the end users are overlooked, in particular, "those historically and systemically marginalised by technology" (Wickerson, 2023). Thus, Co-Design in engineering can be used to address racism, sexism, ableism, and change the systematic privilege of able white men demographics in power and provide justice to marginalised populations who tend to receive inappropriate medical services (Wickerson, 2023).

Co-Design in Tladi, Mokgohloa and Bignotti (Tladi et al., 2021) paper, published in a journal of Industrial Engineering, expands on the necessity for "diversification of paradigms that aim to highlight the significance of contextualised local knowledge". They state,

> It has been noted that the dominance of Western knowledge—particularly in the field of development—has resulted in the marginalisation and disqualification of non-Western knowledge systems. This has led to an acknowledgement that more participatory voices are required, and that the indigenous knowledge of the involved parties (i.e., the tacit knowledge of the beneficiaries) is required to increase participation.
>
> (Tladi et al., 2021)

Recognising the power balance and bottom-up approach is clear from the discussions; however, it seems that in the selected engineering papers, the wickedness of the situations has not been explored and identified as it did in the relevant literature of design discussions in recent years.

10.10 CONCLUSION

While the formalisation of Co-Design processes reduced the emphasis on intuition and shared the power of designers, it did not eliminate the designers' creative tasks. In Co-design practice, both in engineering and design the focus has been shifted from creative thinking around things towards the design of collaboration techniques, processes, and tools. The recent efforts facilitate the co-design research into best practices and the development of new processes.

Co-Design is implementing pragmatic stages to engage stakeholders, matching their experiences, motivations, and roles. By bringing together diverse perspectives, it ensures that the design process is inclusive and considers the needs and insights of all parties involved.

Co-Design today demonstrates interdisciplinarity and bridges knowledge, skills, and practices, from sociology and ethnography to robotics and augmented reality. As demonstrated through this review of literature, it has expanded the boundaries of disciplines, specifically in engineering.

In summary, Co-Design represents a powerful and versatile approach to design that leverages the strengths of both practical processes and philosophical methodologies. It not only aims to produce better design outcomes but also seeks to redefine the way design is practiced, fostering a more inclusive, collaborative, and equitable design culture.

REFERENCES

Aboutajedyne, I., Jordan, S. S., & Aboutajeddine, A. (2022). *Work in progress: Using community-based participatory design and a context canvas to design engineering design courses.* ASEE Annual Conference and Exposition.

Albadra, D., Elamin, Z., Adeyeye, K., Polychronaki, E., Coley, D. A., Holley, J., & Copping, A. (2021). Participatory design in refugee camps: Comparison of different methods and visualization tools. *Building Research & Information*, 49(2), 248–264. https://doi.org/10.1080/09613218.2020.1740578

Bazzano, A. N., Martin, J., Hicks, E., Faughnan, M., & Murphy, L. (2017). Human-centred design in global health: A scoping review of applications and contexts. *PLoS One*, 12(11), e0186744. https://doi.org/10.1371/journal.pone.0186744

Beck, E. E. (2002), P for political: Participation is not enough. *Scandinavian Journal of Information Systems*, 14(1). https://doi.org/https://aisel.aisnet.org/sjis/vol14/iss1/1

Ben Guefrech, F., Boujut, J. F., Dekoninck, E., & Cascini, G. (2023). Studying interaction density in co-design sessions involving spatial augmented reality. *Research in Engineering Design*, 34(2), 201–220. https://doi.org/10.1007/s00163-022-00402-2

Björgvinsson, E., Ehn, P., & Hillgren, P.-A. (2010). Participatory design and "democratizing innovation". *Proceedings of the 11th Biennial Participatory Design Conference*, Sydney, Australia. https://doi.org/10.1145/1900441.1900448

Björgvinsson, E., Ehn, P., & Hillgren, P.-A. (2012). Design things and design thinking: Contemporary participatory design challenges. *Design Issues*, 28(3), 101–116.

Brysch, S., Gruis, V., & Czischke, D. (2023). Sharing is saving? Building costs simulation of collaborative and mainstream housing designs. *Buildings*, 13(3), 821. https://www.mdpi.com/2075-5309/13/3/821

Buchanan, R. (2001). Design research and the new learning. *Design Issues*, 17(4), 3–23.

Busch, O., & Palmås, K. (2023). *The corruption of co-design: Political and social conflicts in participatory design thinking.* Routledge, Taylor & Francis Group.

Büscher, M., Christensen, M., Hansen, K. M., Mogensen, P., & Shapiro, D. (2009). Bottom-up, top-down? Connecting software architecture design with use. In A. Voss, M. Hartswood & R. Procter (Eds.), *Configuring user-designer relations: Interdisciplinary perspectives.* Springer-Verlag.

Casiddu, N., Burlando, F., Porfirione, C., & Vacanti, A. (2022). Co-designed social robotic system in Si-robotics project. *Ambient Assisted Living* (Vol. 884). Springer. https://doi.org/doi:10.1007/978-3-031-08838-4_13

Collina, L., Galluzzo, L., Borin, A., & Mastrantoni, C. (2022). Co-designing spaces for public service. *Designs*, 6(2), 26. https://www.mdpi.com/2411-9660/6/2/26

Corrado, S., Santopietro, L., Annunziata, A., Piro, R., Gatto, R. V., Scorzelli, R., Rahmani, S., Scorza, F., & Murgante, B. (2024). Strategies for democratizing development. Application of Geodesign in a low-context culture. *Innovation in Urban and Regional Planning*. https://doi.org/doi:10.1007/978-3-031-54118-6_14

Cross, N. (1982). Designerly ways of knowing. *Design Studies*, 3(4), 221–227. https://doi.org/10.1016/0142-694X(82)90040-0

Cross, N. (2011). *Design thinking: Understanding how designers think and work.* Berg Publishers. Table of contents only. http://www.loc.gov/catdir/toc/fy11pdf03/2010053657.html

de la Guía, L. S., Puyuelo Cazorla, M., & de-Miguel-Molina, B. (2017). Terms and meanings of "participation" in product design: From "user involvement" to "co-design. *The Design Journal, 20*(sup1), S4539–S4551.

De Medici, S., Marchiano, G., & Pinto, M. R. (2023). A participatory project for the Librino SociaHousing Community. *Vitruvio, 8*(1), 100–113. https://doi.org/doi:10.4995/vitruvio-ijats.2023.19495

Dong, J., Kuo, J., Chen, P., & Bachman, J. (2022). The design of interactive video modules using asset-based participatory design thinking to increase student engagement in engineering. ASEE Annual Conference & Exposition, Minneapolis.

Dortheimer, J., Yang, S., Yang, Q., & Sprecher, A. (2023). Conceptual architectural design at scale: A case study of community participation using crowdsourcing. *Buildings, 13*(1), 222. https://doi.org/10.3390/buildings13010222

Ehn, P. (2008). Participation in design things. Proceedings of the Tenth Anniversary Conference on Participatory Design 2008, Bloomington, Indiana.

Ehab, A., Burnett, G., & Heath, T. (2023). Enhancing public engagement in architectural design: A comparative analysis of advanced virtual reality approaches in building information modeling and gamification techniques. *Buildings*, 13(5), 1262. https://www.mdpi.com/2075-5309/13/5/1262

Enrique, V., Lorena, R., & Cynthia, A. (2016). The technology transfer systems in communities, product versus processes, *Procedia Engineering, 145*, 364–371. https://doi.org/10.1016/j.proeng.2016.04.091

Flatscher, M., & Riel, A. (2016). Stakeholder integration for the successful product–process co-design for next-generation manufacturing technologies, *CIRP Annals, 65*(1), 181–184. https://doi.org/10.1016/j.cirp.2016.04.055

Francis, K., & Murtha, S. (2023). The inclusion and efficacy of first-person narrative in the design of long-term care homes, *Architectural Science Review, 66*(5), 360–371. https://doi.org/10.1080/00038628.2021.1917336

Gaver, B., Dunne, T., & Pacenti, E. (1999). Design: Cultural probes. *Interactions, 6*(1), 21–29.

Govender, V., & Loggia, C. (2023). Adaptive co-design for self-help housing in Durban. *Construction in 5D: Deconstruction, Digitalization, Disruption, Disaster, Development.* https://doi.org/doi:10.1007/978-3-030-97748-1_4

Gray, C., & Malins, J. (2004). *Visualizing research: A guide to the research process in art and design.* Ashgate.

Hendriks, N., Slegers, K., & Duysburgh, P. (2015). Codesign with people living with cognitive or sensory impairments: A case for method stories and uniqueness, *CoDesign, 11*(1), 70–82. https://doi.org/10.1080/15710882.2015.1020316

Hirsch, T. (2016). Surreptitious communication design. *Design Issues, 32*(2), 64–77.

IDEO.org (2015). *The Field Guide to Human-Centered Design.* https://design-kit-production.s3.us-west-1.amazonaws.com/Field_Guides/Field+Guide+to+Human-Centered+Design_IDEOorg_English.pdf?utf8=%E2%9C%93&_method=patch&authenticity_token=QZRbnzBBPY3M%2FCd3xeDx424iAXgVkgcTAi74f6cW4pU%3D&resource%5Btitle%5D=&resource%5Bsubtitle%5D=&resource%5Bauthor%5D=&resource%5Babout%5D=

Jagtap, S. (2022a). Co-design with marginalised people: designers' perceptions of barriers and enablers. *CoDesign: International Journal of CoCreation in Design and the Arts*, 18. https://doi.org/10.1080/15710882.2021.1883065

Jagtap, S. (2022b). Codesign in resource-limited societies: Theoretical perspectives, inputs, outputs and influencing factors. *Research in Engineering Design*, *33*(2), 191–211. https://doi.org/10.1007/s00163-022-00384-1

Jiang, B. (2019). Christopher Alexander and his life's work: The nature of order. *Urban Science*, 3(1), 30. https://www.mdpi.com/2413-8851/3/1/30

Kwieciński, K., & Słyk, J. (2023). Interactive generative system supporting participatory house design. *Automation in Construction*, *145*, Article 104665. https://doi.org/10.1016/j.autcon.2022.104665

Leonard, D., & Rayport, J. (1997). Spark innovation through empathic design. *Harvard Business Review*, *75*, 102–113. https://doi.org/10.1142/9789814295505_0016

Li, N., Kramer, J., Gordon, P., & Agogino, A. (2018). Co-author network analysis of human-centered design for development, *Design Science*, *4*, e10, Article e10. https://doi.org/10.1017/dsj.2018.1

Liu, C., dos Santos Gonçalves, J., & Quist, W. (2023). Participation as a tool for the sustainable redesign of vacant heritage: The case of Politiebureau Groningen Centrum. *Buildings*, *13*(2), Article 515. https://doi.org/10.3390/buildings13020515

Liu, Z. (2022). Housingprime App, A Participatory Design Platform for Housing Customization and Community Self-organization. Proceedings of the 7th International Conference on Architecture, Materials and Construction, Cham. https://doi.org/doi:10.1007/978-3-030-94514-5_11

Low, J. (2019). Design is political: White supremacy and landscape urbanism, *Agora Journal of Urban Planning and Design*, 126–136.

Lubis, P. Y., Shahri, B., & Ramirez, M. (2024). Fostering community empowerment: A human-centered approach to designing clean water solutions in a Jakarta slum. *Journal of Developing Societies*, 40(1), 7–26. https://doi.org/10.1177/01697 96x231222058

Madden, D., Cadet-James, Y., Atkinson, I., & Watkin Lui, F. (2014). Probes and prototypes: A participatory action research approach to codesign, *CoDesign*, *10*(1), 31–45. https://doi.org/10.1080/15710882.2014.881884

Matthews, B., Doherty, S., Worthy, P., & Reid, J. (2023). Design thinking, wicked problems and institutioning change: A case study, *CoDesign*, *19*(3), 177–193. https://doi.org/10.1080/15710882.2022.2034885.

Meinel, C., Leifer, L., & Plattner, H. (2011). *Design Thinking: Understand-Improve-Apply*. Springer. https://doi.org/10.1007/978-3-642-13757-0

Meroni, A., Selloni, D., & Rossi, M. (2018). *Massive codesign: A proposal for a collaborative design framework*. FrancoAngeli.

Miranda, C., & McClam, R. (2023). Work in progress: Using participatory design and qualitative research strategies in the development of a new faculty mentoring program for undergraduate engineering students. *2023 ASEE Annual Conference & Exposition*, Baltimore, Maryland.

Miyoshi, K. (2021). *Designing objects in motion: Exploring kinaesthetic empathy*. Birkhäuser. https://doi.org/10.1515/9783035621105

Moggridge, B., & Atkinson, B. (2007). *Designing interactions* (Vol. 17). MIT press Cambridge.

Moher, D., Liberati, A., Tetzlaff, J., & Altman, D. G.; PRISMA Group (2009). Preferred reporting items for systematic reviews and meta-analyses: The PRISMA statement. *Annals of Internal Medicine*, *151*(4), 264–269.

Molnar, S., & Palmås, K. (2022). Dissonance and diplomacy: Coordination of conflicting values in urban co-design. *CoDesign*, *18*(4), 416–430. https://doi.org/10.1080/15710882.2021.1968441

Ouzzani, M., Hammady, H., Fedorowicz, Z., & Elmagarmid, A. (2016). Rayyan: A web and mobile app for systematic reviews, *Systematic Reviews, 5*. https://doi.org/10.1186/s13643-016-0384-4

Parsons, G. (2016). *The Philosophy of Design*. Polity.

Plank, A. (2016). The hidden risk in user-generated content: An investigation of ski tourers' revealed risk-taking behavior on an online outdoor sports platform. *Tourism Management, 55*, 289–296. https://doi.org/10.1016/j.tourman.2016.02.013

Postma, C., Zwartkruis-Pelgrim, E., Daemen, E., & Du, J. (2012). Challenges of doing empathic design: Experiences from industry. *International Journal of Design, 6*(1).

Rajapakse, R., Brereton, M., & Sitbon, L. (2021). A respectful design approach to facilitate codesign with people with cognitive or sensory impairments and makers. *CoDesign, 17*(2), 159–187. https://doi.org/10.1080/15710882.2019.1612442

Ramstad, I., & Glesaaen, P., & Martina, K. (2020). Participatory disaster risk reduction in developing countries. https://doi.org/10.35199/NORDDESIGN2020.36

Ranabahu, N., Shahri, B., Reid, K., & Hapuku, A. (2024). *How to Engage with Community Advisory Groups for Research Studies*. https://doi.org/10.4135/9781529683097

Ribeiro, J., & Roque, L. (2022). Playful probing: Towards understanding the interaction with machine learning in the design of maintenance planning tools. *Aerospace, 9*(12). https://doi.org/doi:10.3390/aerospace9120754

Rittel, H. W. J., & Webber, M. M. (1973). Dilemmas in a general theory of planning. *Policy Sciences, 4*(2), 155–169. https://doi.org/10.1007/BF01405730

Rodrigues, A. S. L., Martinez, L. B. A., & Silveira, Z. C. (2023). Guidelines for user requirements elicitation in design for assistive technology: A shower chair case study. *Procedia CIRP, 119*, 121–126. https://doi.org/doi:10.1016/j.procir.2023.02.128

Ruiz, P. P., & Maier, A. M. (2012). *Towards describing co-design by the integration of engineering design and technology and innovation management literature*. NordDesign 2012, Aalborg, Denmark.

Saeidi, S., Anderson, M. D., & Davidová, M. (2023). Kindness in architecture: The multispecies co-living and co-design, *Buildings, 13*(8), 1931. https://www.mdpi.com/2075-5309/13/8/1931

Sanders, E. B. N. (2006). Design research in 2006. *Design Research Quarterly, 1*(1), 1–8.

Sanders, E. B. N., & Stappers, P. J. (2008). Co-creation and the new landscapes of design. *Co-Design, 4*(1), 5–18. https://doi.org/10.1080/15710880701875068

Sanders, E. B. N., & Stappers, P. J. (2014). Probes, toolkits and prototypes: Three approaches to making in codesigning. *CoDesign, 10*(1), 5–14. https://doi.org/10.1080/15710882.2014.888183

Shapiro, D. (2005). Participatory design: the will to succeed. *Proceedings of the 4th Decennial Conference on Critical Computing: Between Sense and Sensibility, Aarhus, Denmark*. https://doi.org/10.1145/1094562.1094567

Sharmin, T., & Khalid, R. (2022). Post occupancy and participatory design evaluation of a marginalized low-income settlement in Ahmedabad, India, *Building Research & Information, 50*(5), 574–594. https://doi.org/10.1080/09613218.2021.2018286

Skoury, L., Treml, S., Opgenorth, N., Amtsberg, F., Wagner, H. J., Menges, A., & Wortmann, T. (2024). Towards data-informed co-design in digital fabrication, *Automation in Construction, 158*, 105229. https://doi.org/10.1016/j.autcon.2023.105229

Smith, J., Lucena, J., & Rivera, A. (2022). Humanitarian engineering, global sociotechnical competency, and student confidence: A comparison of in-person, virtual, and hybrid learning environments. *ASEE Annual Conference & Exposition*, Minneapolis.

Spinuzzi, C. (2005). The methodology of participatory design, *Technical Communication, 52*, 163–174.

Star, S. L. (2010). This is not a boundary object: Reflections on the origin of a concept, *Science, Technology, & Human Values*, *35*(5), 601–617. http://www.jstor.org/stable/25746386

St John, N., & Akama, Y. (2022). Reimagining co-design on country as a relational and transformational practice, *CoDesign*, *18*(1), 16–31. https://doi.org/10.1080/15710882.2021.2001536

Tladi, B., Mokgohloa, K., & Bignotti, A. (2021). Towards a relational framework of design. *The South African Journal of Industrial Engineering*, *32*(3). https://doi.org/doi:10.7166/32-3-2622

von Busch, O., & Palmås, K. (2023). *The Corruption of Co-Design: Political and Social Conflicts in Participatory Design Thinking*. Routledge. https://doi.org/10.4324/9781003281443

von Hippel, E. (2005). *Democratizing innovation*. MIT Press. http://www.loc.gov/catdir/toc/fy053/2004061060.html

Voss, A., Hartswood, M., Procter, R., Rouncefield, M., Slack, R. S., & Buscher, M. (2009). *Configuring user-designer relations: Interdisciplinary perspectives*. Springer. https://doi.org/10.1007/978-1-84628-925-5

Wickerson, G. (2023). Community-Driven, Participatory Engineering Design Frameworks to Shape Just, Liberatory Health Futures. *ASEE Annual Conference and Exposition, Conference Proceedings*.

Wilson, S., Roper, A., Marshall, J., Galliers, J., Devane, N., Booth, T., & Woolf, C. (2015). Codesign for people with aphasia through tangible design languages. *CoDesign*, *11*(1), 21–34. https://doi.org/10.1080/15710882.2014.997744

APPENDIX 1

This is the complete search query and filtering:

TITLE-ABS-KEY (codesign OR co-design OR participatory AND design OR collaborative AND design) AND (LIMIT-TO (PUBSTAGE, "final")) AND (LIMIT-TO (SUBJAREA, "ENGI") OR EXCLUDE (SUBJAREA, "COMP") OR EXCLUDE (SUBJAREA, "PHYS") OR EXCLUDE (SUBJAREA, "MATE") OR EXCLUDE (SUBJAREA, "MATH") OR EXCLUDE (SUBJAREA, "SOCI") OR EXCLUDE (SUBJAREA, "ENER") OR EXCLUDE (SUBJAREA, "EART") OR EXCLUDE (SUBJAREA, "BUSI") OR EXCLUDE (SUBJAREA, "CENG") OR EXCLUDE (SUBJAREA, "DECI") OR EXCLUDE (SUBJAREA, "ENVI") OR EXCLUDE (SUBJAREA, "MEDI") OR EXCLUDE (SUBJAREA, "HEAL") OR EXCLUDE (SUBJAREA, "BIOC") OR EXCLUDE (SUBJAREA, "CHEM") OR EXCLUDE (SUBJAREA, "AGRI") OR EXCLUDE (SUBJAREA, "ARTS") OR EXCLUDE (SUBJAREA, "ECON") OR EXCLUDE (SUBJAREA, "NURS")) AND (EXCLUDE (DOCTYPE, "cr") OR EXCLUDE (DOCTYPE, "bk") OR EXCLUDE (DOCTYPE, "ch") OR EXCLUDE (DOCTYPE, "re") OR EXCLUDE (DOCTYPE, "ed") OR EXCLUDE (DOCTYPE, "er")) AND (LIMIT-TO (LANGUAGE, "English") OR EXCLUDE (LANGUAGE, "Italian") OR EXCLUDE (LANGUAGE, "Polish") OR EXCLUDE (LANGUAGE, "Japanese") OR EXCLUDE (LANGUAGE, "Swedish")) AND (LIMIT-TO (EXACTKEYWORD,

"Co-designs") OR LIMIT-TO (EXACTKEYWORD, "Participatory Design") OR LIMIT-TO (EXACTKEYWORD, "Co-design") OR EXCLUDE (EXACTKEYWORD, "Hardware Software Codesign") OR EXCLUDE (EXACTKEYWORD, "HW/SW Codesign") OR EXCLUDE (EXACTKEYWORD, "Hardware-software Codesign") OR EXCLUDE (EXACTKEYWORD, "Computer Hardware") OR EXCLUDE (EXACTKEYWORD, "Computer Software") OR EXCLUDE (EXACTKEYWORD, "Hardware And Software") OR EXCLUDE (EXACTKEYWORD, "Hardware/software Codesign") OR EXCLUDE (EXACTKEYWORD, "HW/SW Co-design") OR EXCLUDE (EXACTKEYWORD, "Hardware-software Co-design") OR EXCLUDE (EXACTKEYWORD, "Hardware/software Partitioning") OR EXCLUDE (EXACTKEYWORD, "Embedded Systems") OR EXCLUDE (EXACTKEYWORD, "CMOS Integrated Circuits") OR EXCLUDE (EXACTKEYWORD, "Controllers") OR EXCLUDE (EXACTKEYWORD, "Field Programmable Gate Arrays (FPGA)") OR EXCLUDE (EXACTKEYWORD, "Algorithms") OR EXCLUDE (EXACTKEYWORD, "Hardware/software Co-design") OR EXCLUDE (EXACTKEYWORD, "Wireless Telecommunication Systems") OR EXCLUDE (EXACTKEYWORD, "Machine Learning"))

APPENDIX 2

Included articles in the systematic review imported from Rayyan in APA7 format

Albadra, D., Elamin, Z., Adeyeye, K., Polychronaki, E., Coley, D. A., Holley, J., & Copping, A. (2021). Participatory design in refugee camps: Comparison of different methods and visualization tools. *Building Research & Information*, 49(2), 248–264. https://doi.org/10.1080/09613218.2020.1740578

Albers, A., Lanza, G., Klippert, M., Schäfer, L., Frey, A., Hellweg, F., Müller-Welt, P., Schöck, M., Krahe, C., Nowoseltschenko, K., & Rapp, S. (2022). Product-Production-CoDesign: An approach on integrated product and production engineering across generations and life cycles. *Procidia CIRP*, 109, 167–172. https://doi.org/doi:10.1016/j.procir.2022.05.231

Arif, S. (2020). Designing an engineering computer instructional laboratory: Working with the panopticon. *ASEE Virtual Conference 2020*.

Ben Guefrech, F., Boujut, J. F., Dekoninck, E., & Cascini, G. (2023). Studying interaction density in co-design sessions involving spatial augmented reality. *Research in Engineering Design*, 34(2), 201–220. https://doi.org/10.1007/s00163-022-00402-2

Bettelli, A., Pluchino, O. V., Dainese, P., Campagnaro, G., Narne, V., Paoli, E., Moro, A., & Gamberini, T. (2022). SHIP project: Designing inclusive, accessible, and sustainable urban parks. *Ambient Assisted Living: Lecture Notes in Electrical Engineering* (Vol. 884). Springer. https://doi.org/doi:10.1007/978-3-031-08838-4_11

Casiddu, N., Burlando, F., Porfirione, C., & Vacanti, A. (2022). Co-designed social robotic system in Si-robotics project. *Ambient Assisted Living: Lecture Notes in Electrical Engineering* (Vol. 884). Springer. https://doi.org/10.1007/978-3-031-08838-4_13

Chowdhury, S. & Schnabel, M. A. (2020). Virtual environments as medium for laypeople to communicate and collaborate in urban design. *Architectural Science Review*, 63(5). https://doi.org/doi:10.1080/00038628.2020.1806031

Cobb, P. J., Pan, W. E. M. W., Lou, N. F. C., Hu, V. W. Q., Cheng, X., & Xiao, M. J. (2021). Sharing the past: The library as digital co-design space for intergenerational heritage preservation. *2021 ACM/IEEE Joint Conference on Digital Libraries*, 27–30 September. https://doi.org/doi:10.1109/JCDL52503.2021.00051

Collina, L., Galluzzo, L., Borin, A., & Mastrantoni, C. (2022). Co-designing spaces for public service. *Designs*, *6*(2), 26. https://www.mdpi.com/2411-9660/6/2/26

Corrado, S., Santopietro, L., Annunziata, A., Piro, R., Gatto, R. V., Scorzelli, R., Rahmani, S., Scorza, F., & Murgante, B. (2024). Strategies for democratizing development. Application of Geodesign in a low-context culture. *Innovation in Urban and Regional Planning: Lecture Notes in Civil Engineering* (Vol 467). Springer. https://doi.org/doi:10.1007/978-3-031-54118-6_14

De Medici, S., Marchiano, G., & Pinto, M. R. (2023). A participatory project for the Librino SociaHousing Community. *Vitruvio*, *8*(1), 100–113. https://doi.org/doi:10.4995/vitruvio-ijats.2023.19495

Domènech-Rodríguez, M., Cornadó, C., Vima-Grau, S., Piasek, G., Varela-Conde, A., & Ravetllat Mira, P. J. (2023). Co-design and co-manufacturing: A multidisciplinary approach through small-scale architectural experiences in Barcelona. *Buildings*, *13*(5). https://doi.org/doi:10.3390/buildings13051159

Dortheimer, J., Yang, S., Yang, Q., & Sprecher, A. (2023). Conceptual architectural design at scale: A case study of community participation using crowdsourcing. *Buildings*, *13*(1), 222. https://doi.org/10.3390/buildings13010222

Gamberini, L., Pluchino, B. D., Grippaldi, P., Dainese, A. Z., Campagnaro, G., Zanella, V., Mapelli, A., Mondini, D., Campana, S., Cipolletta, G., Spagnolli, S., & Sozza, A. A. (2022). DOMHO: Internet of Things for ambient assisted co-housing. *Ambient Assisted Living: Lecture Notes in Electrical Engineering* (Vol. 884). Springer. https://doi.org/doi:10.1007/978-3-031-08838-4_8

Govender, V., & Loggia, C. (2023). Adaptive Co-Design for Self-Help Housing in Durban. *Construction in 5D: Deconstruction, Digitalization, Disruption, Disaster, Development*. https://doi.org/doi:10.1007/978-3-030-97748-1_4

Happonen, A., Wolfe, A., & Palacin, V. (2022). From Data Literacy to Co-design Environmental Monitoring Innovations and Civic Action. *Proceeding of 2021 International Conference on Wireless Communications, Networking and Applications*. https://doi.org/doi:10.1007/978-981-19-2456-9_42

Jagtap, S. (2022b). Codesign in resource-limited societies: Theoretical perspectives, inputs, outputs and influencing factors. *Research in Engineering Design*, *33*(2), 191–211. https://doi.org/10.1007/s00163-022-00384-1

Kwieciński, K., & Słyk, J. (2023). Interactive generative system supporting participatory house design. *Automation in Construction*, *145*, Article 104665. https://doi.org/10.1016/j.autcon.2022.104665

Lahti, M. & Nenonen, S. (2021). Design science and co-designing of hybrid workplaces. *Buildings*, 11(3). https://doi.org/doi:10.3390/buildings11030129

Liu, C., dos Santos Gonçalves, J., & Quist, W. (2023). Participation as a tool for the sustainable redesign of vacant heritage: The case of Politiebureau Groningen Centrum. *Buildings*, *13*(2), Article 515. https://doi.org/10.3390/buildings13020515

Liu, Z. (2022). Housingprime App, A Participatory Design Platform for Housing Customization and Community Self-organization. Proceedings of the 7th International Conference on Architecture, Materials and Construction, Cham. https://doi.org/doi:10.1007/978-3-030-94514-5_11

Pelorosso, R., Zingoni, A., Ruko, S., & Calabrò, G. (2024). Map4Accessibility Project, An Inclusive and Participated Planning of Accessible Cities: Overview and First Results. *Innovation in Urban and Regional Planning* (Vol. 463). Springer. https://doi.org/doi:10.1007/978-3-031-54096-7_24

Ramstad, I., & Glesaaen, P., & Martina, K. (2020). Participatory disaster risk reduction in developing countries. https://doi.org/10.35199/NORDDESIGN2020.36

Ribeiro, J., & Roque, L. (2022). Playful probing: Towards understanding the interaction with machine learning in the design of maintenance planning tools. *Aerospace*, *9*(12). https://doi.org/doi:10.3390/aerospace9120754

Rodrigues, A. S. L., Martinez, L. B. A., & Silveira, Z. C. (2023). Guidelines for user requirements elicitation in design for assistive technology: A shower chair case study. *Procedia CIRP*, *119*, 121–126. https://doi.org/doi:10.1016/j.procir.2023.02.128

Scott, P. T., Michinel, G. T., Helton, A., Lacy, D., Walker, H., & Shelke, J. U. (2021). Case study: converting a cracker idled for 15+ years into a profitable process unit.

Sharmin, T., & Khalid, R. (2022). Post occupancy and participatory design evaluation of a marginalized low-income settlement in Ahmedabad, India. *Building Research & Information*, *50*(5), 574–594. https://doi.org/10.1080/09613218.2021.2018286

Tladi, B., Mokgohloa, K., & Bignotti, A. (2021). Towards a relational framework of design. *The South African Journal of Industrial Engineering*, *32*(3). https://doi.org/doi:10.7166/32-3-2622

Wickerson, G. (2023). Community-Driven, Participatory Engineering Design Frameworks to Shape Just, Liberatory Health Futures. *ASEE Annual Conference and Exposition, Conference Proceedings*.

Woolner, P., Cardillino, P. (2021). Crossing contexts: Applying a system for collaborative investigation of school space to inform design decisions in contrasting settings. *Buildings*, 11(11), 496. https://doi.org/doi:10.3390/buildings11110496

11 Design, Thinking, and Metaphors

On the Cognitive Style of Engineering and Design Education

Karl Palmås

11.1 INTRODUCTION

The notion of design thinking has been a topic of discussion at least since the arrival of what Nigel Cross (1984) describes as the third generation of design studies. Thus, scholars have asked what it means to think like a designer, or like a member of the broader category of design professions. However, the discussion on design thinking has been less concerned with the question of how humans – designers and non-designers alike – think with reference to designed artifacts. Conceived in this way, "design thinking" is less a matter of interrogating and mimicking the cognitive styles of practicing designers, and more about exploring how design shapes culture and social imaginaries.

The usual way to attack the issue of design and society is through the study of how design shapes social action, with references to affordances (Norman, 1988) or invitation-inhibition structures (Verbeek, 2005). Such aspects of design may also involve references to how design may have ideological effects, for instance, by promoting consumption (Fry, 1999) or assisting in new ways of governing societies (von Busch and Palmås, 2023). However, by shaping modes of thought, design also participates in shaping public discourse. This has implications for design education.

Before discussing these implications, the chapter will first briefly discuss the stakes of teaching design sciences. It will then introduce German-American political thinker Hannah Arendt, outlining her distinction between the 'work' of making artifacts, and the 'action' of imagination in the public realm. The chapter will then critique her implication that the making of artifacts is a merely instrumentalist pursuit, because – for good and bad – design assists in generating figures of thought. We cannot dissociate the things we think from the things that we build. To this end, the text will survey different accounts of how thinking is tied to metaphoric from designed artifacts. The chapter ends with a return to the issue of design education.

DOI: 10.1201/9781003487524-11

11.2 DESIGN EDUCATION FOR JUDGMENT

In the education for the design sciences, judgment is a crucial term. For instance, there is engineering judgment, which stretches from the most mundane technical choices to more overarching ethical judgments on our uses of technology (Palmås, 2024). This faculty of judgment is similar to the kind of judgment trained in other professional schools – notably business schools. Here, judgment implies being a wise decision-maker, someone who does not only possess book knowledge, but also a capacity to make assessments based on either experience or imagination.

Still, judgment in design or engineering is different from the judgment trained at a business school. At technical universities and in design schools, we deal with technological tools and designed artifacts. We train for "making and fabricating activities" – for the kind of activities that Arendt associates with *homo faber;* the making human.

In preparation for a professional life of making, teachers assign design challenges, case study discussions, games-playing, and scenario-building. Such teaching exercises serve to train judgment by simulating the experiences doing design; the kind of real-life experiences that the students have yet to take part in as design professionals. However, there is also a place for reading classics in social thought. This is partly a matter of merely introducing societal aspects to their education. If you want to understand what it is like to work with technology, you need to know something about sociological aspects of technology, the historical formation of both ideas and innovation, and the relation to the arts. However, this is also a matter of training the mind: As some scholars suggest, broad liberal arts educations train the mind to see connections between different fields of knowledge. (Kallenberg, 2015; Bucciarelli and Drew, 2015)

Still, those of us teaching the design sciences are not really in the same business as those teaching the humanities. As mentioned above, we teach for making and fabricating, and this sets us apart from the humanities. Indeed, if we are to follow Arendt in *The Human Condition*, there is a sharp distinction between the "work" of making, and the "action" created through language.

11.3 MAKING VERSUS ACTING

Published in 1958, in the context of a nascent Cold War and space race, *The Human Condition* is Arendt's major exploration of what it means to be human in a modern, technologically advanced world. It is also considered the most systematic expression of her thought. For Arendt, humans are biological creatures who *labor* for sustenance and are making creatures who *work* by creating artifacts for various uses. Nevertheless, our true humanity comes out in the stories we tell about ourselves and the human experience. It is only through the latter that we are political creatures who engage in the public sphere, through *action*. Thus, the Arendtian account of making activities (work) is one that emphasizes the instrumentality of technology. Technology pins the world down, holding imagination in check, eliminating the element of chance in human history.

This conception of technology can be placed in context. Here, it is productive to consider Marshall Berman's point about Arendt and *The Human Condition*: It is written during a post-WWII moment when modernism in thought had become severed from modernity (Berman, 1982: 310). Earlier modern thinkers like Marx had marveled at the modern world of technology, writing *with* it. The great thinkers of the 1950s, Berman suggests, were either ignoring the paraphernalia of modernity or writing *against* it.

For Arendt, then, the story of the modern world is one where there is a "substitution of making for action" (Arendt, 1958: 229), which has led a "concomitant degradation of politics into a means to obtain an allegedly 'higher' end". This instrumentalist orientation starts already with Plato, for whom this higher end implied "the protection of the good men from the rule of the bad". The tendency then continues into the modern age, in which the higher end of productivity and progress becomes the dominating force.

The main culprit in instigating this degradation of politics in the modern age was philosopher Thomas Hobbes. He was the one who introduced "the new concepts of making and reckoning into political philosophy" (Arendt, 1958: 300). Here, "reckoning" implies anticipation of uncertain futures and chance events – indeed a topic of discussion in design studies and organizational studies today. For Arendt, though, such reckoning is problematic: For the making human, the anticipation of futures is solely a means to maintain the *status quo*, not an exercise in imagining alternative futures. Thus, "processes, therefore, and not ideas, the models and shapes of things to be, become the guide for the making and fabricating activities of *homo faber* in the modern age."

This is, nevertheless, a problematic account of the stakes of making. First, one may question the extent to which making is necessarily tied to instrumentalism and control. Second, there are good reasons for suggesting that "ideas, the models and shapes of things to be" do not exist in some ideal abstract space. Instead, we think through what we see and experience in the world of appearances. The remainder of this text will engage with these two points in tandem.

11.4   MAKING AS ADVENTURE AND AS ROUTINE

First, there is the question of making as instrumentalization and control. Arendt's position is in alignment with much philosophy of technology, but there is more to the story. Here, we can again go back to Marshall Berman (1982) and his distinction between modernization as adventure and modernization as routine. He is doing so in his reading of Fyodor Dostoyevsky's *Notes from the Underground*. The underground man rails against Crystal Palace – the building erected in the context of the 1851 Great Exhibition in London (See Figure 11.1). The man in Dostoyevsky's book saw the building as a symbol of the onset of the new and the rational, which erases the element of chance and wonder in human existence. This, in turn, is a comment on the Russian early socialist writers, who had made the Crystal Palace their chosen utopian model for future civilization.

FIGURE 11.1   Crystal Palace in Sydenham, 1854. (Photographed by Philip Henry Delamotte).

Still, the story is more nuanced than that. We know that Dostoyevsky himself was horrified by seeing Crystal Palace. He saw it as an accomplishment that was so final, so overwhelming that it represented the end of history. On the other hand, Dostoyevsky – and his underground man – was also enamored by engineering. The protagonist in *Notes from the Underground* postulates that:

> Man is preeminently a creative animal, predestined to strive consciously toward a goal, and to engage in engineering; that is, eternally and incessantly, to build new roads, *wherever they may lead.*

Berman notes that "wherever they may lead" is written in italics. Why is this so? Berman suggests that Dostoyevsky is not critical toward technology or engineering as such. The key is to somehow maintain the wonder of technology and not let it settle into a routine. In other words, Dostoyevsky was writing against the early socialists' desire to turn the singular marvel of the Crystal Palace into a stale, standardized, and fixed expression. This is a general problem of technology. As we know from sociologist Andrew Barry, just like the doing of politics can have "anti-political" effects, turning dissensus to consensus, the doing of technological development can have "anti-inventive" outcomes, shutting down the routes to alternative futures (Barry, 2001: 268). Still, it does not have to.

11.5   THINKING IN METAPHORICS

Then, there is the question of whether "ideas, the models and shapes of things to be" exist solely in some abstract space. There is also the related question of whether language and the telling of stories is the thing that truly makes us human. Those of us who are interested in the design sciences will probably disagree. As design scholar John Heskett (2002: 6) points out, it is not only language that makes humans stand out from other animals; it is also the ability to design and make things that define us as human.

Still, let us stay with the social theorists. Arendt distinguishes thinking from judgment precisely on the point of abstraction. Thinking tends toward the abstract, judgment tends toward the concrete. "Thinking deals with invisibles, with representations of things that are absent; judging always concerns particulars and things close at hand" (Arendt, 2003: 189). However, in her later work on thinking, she discusses the metaphorical nature of thought. This is, I believe, where the distinction between work and action starts to unravel. Can we really dissociate the realm of artifacts from the realm of language-based communication and action?

For one, we may turn to Lakoff and Johnson's classic account in *Metaphors we Live By*. In the 2003 preface to their original 1980 monograph, they are at pains to emphasize that metaphor is not about language. They stress that metaphors are conceptual, and that we think through the cross-mapping different conceptual domains. They ask: "Are those metaphorical mappings purely abstract and arbitrary? The empirical answer is 'no'. They are shaped and constrained by our bodily experiences in the world" (Lakoff and Johnson, 2003: 247). We think with reference to what we experience. In this way, conceptual constructions can be abstracted from actual constructions. Heskett suggests:

> Abstraction enables capacities to be separated from specific problems, to be generalized, and flexibly adapted to other problems. [...] In other words, objects are not just expressions of a solution to a particular problem at any point in time, but can extend much further, into embodying ideas [...]
>
> (Heskett, 2002: 12)

This idea is present in the work of social theorist Peter Sloterdijk, notably his trilogy on spheres. As Stuart Elden and Eduardo Mendieta point out:

> In Sloterdijk's work we find a continuous play among image, imagination, and imaginary that shuttles back and forth between what we experience and see, and what we can imagine or cannot imagine because we have not seen an image of what it could be like.
>
> (Elden and Mendieta, 2009: 6)

A similar story can be found in the work of philosopher of science Michel Serres, who highlights the ways in which history has implied a "the parallel development of scientific, philosophical, and literary trends" (Serres, 1982: xi).

Serres' paradigmatic example is the relation between the steam engine, the science of thermodynamics, and the social theory of the mid- to late-1800s. In this story, engineer Sadi Carnot plays a crucial role in defining the engine as a device that exploits the heat difference between reservoirs of hot and cold. This abstract idea of a device that evens out differences is then adopted in several conceptual domains, such as social theory, psychology, and literature.

All of these accounts suggest that we think with reference to designed artifacts, and perhaps visual perception of such artifacts. This implies that there may be a "cognitive style" that is associated with design, hence "design thinking". However, if designers and engineers – representatives of *homo faber* – really have this impact on the world, what does that mean for education?

11.6 EDUCATING FOR MAKING AND IMAGINATION

If the design sciences are not just shaping the physical world, but also mental worlds, there are (at least) two issues that emerge for education. First, how can these stakes of design be explicated to students? Second, how can educators encourage design and technological development that encourages imagination and not instrumentalization, adventure and not routine?

The question of how to explicate the stakes of design does, to some degree, imply highlighting social and other contextual aspects of the design sciences. However, as hinted above, this is not merely about teaching context for context's sake. Liberal arts elements blended into engineering education, as proposed by Bucciarelli and Drew (2015), has the benefit of fostering a particular cognitive style. This style implies being capable of identifying connections across different knowledge domains (Kallenberg, 2015), and of recognizing how philosophy, history, literature, and engineering science are "conjoined". (Bucciarelli and Drew, 2015: 111) Such an approach to engineering education demonstrates how making and public imagination is – contrary to Arendt's position – intertwined. It also trains the kind of abstraction that Heskett associates with design.

The second question implies taking seriously the fact that making somehow informs public imagination. How can educators train practitioners whose work emphasizes the creative and imaginative – not the instrumentalist and routine? Here we can turn to George Orwell. He may be an unlikely candidate in a text that critiques the proposition that thinking is purely a matter of language. Indeed, the novel *1984* is sometimes read as a story about controlling thought through controlling language. Yet, in the end of the novel, we learn that the introduction of "Newspeak" turned out to be a failure for the government. This could be read as Orwell suggesting that there are limits of controlling minds through language.

For the purposes of this text, it is productive to look elsewhere in his work, specifically his suggestions for writing in the essay "Politics and the English language" (Orwell, 2013). In it, he is in alignment with the idea of thinking in metaphors, which to some degree are visual. However, his main point is that metaphors can both liberate and stultify thought. Stale and standardized metaphors will render your thoughts unoriginal: "By using stale metaphors, similes, and idioms, you save much mental effort, at the cost of leaving your meaning vague, not only for your reader but for yourself." Further, he suggests that when we are using ready-made metaphors in an unreflective way, we are actually not thinking. Arendt (1968) also makes a similar point, suggesting that the use of mental clichés in the public sphere is a sign of "dark times".

If it is the case that our thinking is conditioned by metaphors borrowed from the world of designed artifacts, and that stale metaphors stultifies thought, there is a case to be made for design and engineering education that habitually challenges the cliché and ready-made. While designers must be realists who make sure that good intentions do not lead to bad outcomes (von Busch and Palmås, 2023), it is also important for them to recognize that the results of their work will become conversation pieces that shape culture and society. Creativity is not solely a means to generate commercializable innovations. In challenging the routine, designers and engineers – as representatives of *homo faber* – expand the shared imagination of the general public.

REFERENCES

Arendt, Hannah (1958) *The Human Condition*. Chicago: The University of Chicago Press.
Arendt, Hannah (1968) *Men in Dark Times*. New York: Harcourt Brace.
Arendt, Hannah (2003) *Responsibility and Judgment*. New York: Schocken Books.
Barry, Andrew (2001) *Political Machines: Governing a Technological Society*. London: Athlone.
Berman, Marshall (1982) *All That Is Solid Melts into Air: The Experience of Modernity*. New York: Simon and Schuster.
Bucciarelli, Louis, and David Drew (2015) "Liberal Studies in Engineering: A Design Plan." *Engineering Studies* 7, no. 2-3: 103–122.
von Busch, Otto, and Karl Palmås (2023) *The Corruption of Co-Design Political and Social Conflicts in Participatory Design Thinking*. London: Routledge.
Cross, Nigel (1984) *Developments in Design Methodology*. Chichester: Wiley.
Elden, Stuart, and Eduardo Mendieta (2009) "Being-with as Making Worlds: the 'second coming' of Peter Sloterdijk." *Environment and Planning D: Society and Space* 27: 1–11.
Fry, Tony (1999) *Defuturing: A New Design Philosophy*. London: Bloomsbury.
Heskett, John (2002) *Design: A Very Short Introduction*. Oxford: Oxford University Press.
Kallenberg, Brad (2015) "Liberal Arts Is More Than 'Perspective." *Engineering Studies* 7, no. 2-3: 132–134.
Lakoff, G., and M Johnson (2003 [1980]) *Metaphors We Live By*. Chicago: University of Chicago Press.
Norman, D. A (1988) *The Psychology of Everyday Things*. New York: Basic Books.
Orwell, George (2013) *Politics and the English Language*. London: Penguin Classics.
Palmås, Karl (2024) "Engineering Judgment and Education: An Arendtian Account", *Engineering Studies*, pp. 1–22.
Serres, Michel (1982) *Hermes: Literature, Science, Philosophy*. Baltimore: The Johns Hopkins University Press.
Verbeek, Peter-Paul (2005) *What Things Do: Philosophical Reflections on Technology, Agency, and Design*. University Park, PA: Penn State University Press.

12 Does Design Thinking Work in the Long-term? A Case Study of Interdisciplinary Research on Aging

P.J. White, Audrey Patocs, Marla Beauchamp, and Parminder Raina

12.1 INTRODUCTION

In the introduction stage of this chapter, we set the context for the research undertaken. We outline the historical evolution of design thinking (DT) and present a backdrop to its use, its perceived shortcomings, its potential future, and gaps in understanding its future. Design as interdisciplinary thinking and DT for interdisciplinary research are then discussed through their shared values. We conclude the introduction section by offering ten summary points from the literature and outline the primary research question: Does DT work in the long term? Through this question we aim to understand the conditions of embedding long-term DT and offer recommendations for future researchers and practitioners to optimize it in long-term projects. We follow the introduction section with the research case study in the second section of the chapter.

12.2 EVOLUTION OF DESIGN THINKING

The last century has seen the field of design evolve extensively both as a practice and a thought process. The commencement of this evolution can be plotted from the early Bauhaus doctrine of the 1920s underpinning the philosophical understanding of design across creative disciplines. From this time, design has been moving in and out of constructivist and positivist paradigms to evolve, understand, and formalize methods and processes (Bousbacı, 2008).

Post World War II, with the application of scientific and technological advancements, saw the development in understanding design as a thought process for addressing complex problems. The publication of Herbert Simon's book *The Sciences of the Artificial* marked an important milestone in this regard, differentiating design as a problem-solving activity (Simon, 1968). Echoing these virtues were contemporaries, Horst Rittel and Melvin Webber. In their paper, "Dilemmas in a General Theory of Planning," they looked at issues within social policy and argued that 'scientific' approaches alone could not resolve complex 'wicked' societal problems (Rittel & Webber, 1973).

From the 1980s there was continuous exploration into the thought processes of design practice and the democratization of design as an interdisciplinary research approach. The publication of Donald Schön's book, *The Reflective Practitioner*, in the early 1980s (Schön, 1983) coincided with the growth of ubiquitous computing in Silicon Valley. This brought together different disciplines, values, and methods, namely design, anthropology, business, and engineering, and created the interdisciplinary nature of participatory-based or user-centered design into the 1990s (Blomberg et al., 2017; Wolff, 2023).

12.3   USE OF DESIGN THINKING AND PERCEIVED SHORTCOMINGS

The use of design to gain economic advantage within business emerged in the United Kingdom and Europe in the 1980s and 1990s (Cross, 2023b). In the United States, design for economic growth continued and consolidated its development into the 2000s with the popularization of DT (Martin, 2009). This was helped by IDEO and Stanford's D School commoditizing design methods for organizations throughout the 2000s and 2010s (Ackermann, 2023). The use of DT quickly grew within business organizations as a problem-solving methodology for innovation, and for finding pathways for new products and services (Brown, 2009). DT extended its reach to redesigning organizations to operate more efficiently, physically, and structurally, and used inwardly in creating more efficient, lean, and adaptive working environments allowing for innovation to occur (Dunne, Eriksson and Kietzmann, 2022; Gaynor et al., 2018). DT's outward use grew in designing customer experiences and understanding human behaviours, and was used to create behaviour change and help lower risks and costs in businesses (Liedtka, 2018). Used more holistically, DT continues to be applied to address diverse complex issues from politics, to education, to health and national public transport systems (White and Kennedy, 2021; Ku and Lupton, 2022; Kennedy, White and Dempsey, 2024).

DT, however, has had its detractors, a shared criticism of DT is in its reach and short-term focus. In this regard, Nussbaum describes it as a "failed experiment" (2011). Iskander has criticized DT's reach and depth to limiting participation in the design process, as a result, being "fundamentally conservative and helping preserve the status quo" (2018). DTs reach is also questioned by Cross, stating that "… thinking and acting are certainly relevant to tackling a broad range of problems, but they are not a universal issue-resolving cure-all" (Cross, 2023b). Ackermann states that "…critics have argued that its short-term focus on novel and naive ideas has resulted in unrealistic and ungrounded recommendations"

(2023). In its defence, DTs depth and reach can be attributed to its corporate background and evolution with the speed and short-term nature of industrial projects. In addition, with a lack of long-term focus on DT, it can be difficult to assess its effectiveness over time. This is what we would like to address in this chapter.

12.4   NEXT PHASE DESIGN THINKING: WHERE NEXT?

According to Cross, DT is now at a transitionary stage, stating that it is possible that "...another version of design thinking as a way of acting within complex, problematic issues may be emerging" (2023b, p. 8). So, the question needs to be asked... Where to go next? What can we learn from the last two decades of DT, its potential and its shortcomings?

Despite these two decades of development, Verganti, Dell'Era and Swan (2021) state that there is still a lack of understanding of whether, why, and when DT contributes to innovation. Therefore, to understand its future, there is a real need to reflect and report on past case studies of its long-term application. However, DT in long-term projects has rarely been reported and or reflected on. Most studies on long-term DT in organizations focus on implementation, and insights within are invaluable in understanding its potential future. For example, Wrigley, Nusem and Straker (2020) outline 7 organizational case studies seeking to integrate design within 12–24 month periods and outline 4 organizational conditions required for DT to attain a long-term impact: strategic vision, facilities, cultural capital, and directives. Using a longitudinal case study approach to understand when and how to implement DT in the innovation process, Cai, Lin and Zhang (2023) found that implementation varies with the phase of innovation and that flexible deployment of DT practices is required as phases progress. In implementing DT to engage communities in Health Innovation, van der Westhuizen et al. (2020), conclude that it offers opportunities for enriching community engagements. However, they note the 'fast, dynamic style of design thinking' is not entirely suited for developing the level of trust and rapport that is required. In response to this, they suggest that 'specific tool kits' need to be created to meet the needs of engagements. Shrivastav and Joshi (2021) discuss the long-term scaling of DT processes to different sectors for new product development. The results of this allowed an organization to capture a significant market share within two years of a new product launch (Shrivastav and Joshi, 2021). We can conclude from the literature above that not withholding DT's strengths, there is a lack of understanding of the long-term use of DT and the specific conditions that allow it to strive.

12.5   DESIGN AS INTERDISCIPLINARY THINKING

The historical evolution of DT shows that its methods and processes are derived from interdisciplinary spaces, e.g., anthropology, business, and engineering. This interdisciplinary composite is continually evolving and emerging. Crossovers are reciprocal, with disciplines that appropriate DT benefiting from the creative, adaptive, and human-centric qualities of its processes. An important foundational quality of DT is including people in the process of change and ensuring the needs of users are expressed (Liedtka, 2018). The design process, by its very nature, is a human-centred and social activity, involving communicating and gaining human insight between disciplinary boundaries and drawing on multi-modal methods is part of the design process. DT regularly requires interdisciplinary involvement; therefore, its process is required to be inclusive across disciplinary domains. Cross (2023a) maintains that the act of designing is something that everyone can do at a certain level and that it is inherent in human cognition and decision-making. Therefore, to create interdisciplinary working environments the inherent design skills within groups need to be harnessed and encouraged.

12.6   DESIGN THINKING FOR INTERDISCIPLINARY RESEARCH

Interdisciplinary research is increasingly regarded as the key to addressing challenges that face contemporary society today (Sun et al., 2021). According to Stember (1991) the problems of the world are not organized according to academic disciplines and complex modern problems such as climate change and resource security are not responsive to single-discipline investigation (Rylance, 2015). Nurturing interdisciplinary research collaborations can increase personal and team achievements with interdisciplinary researchers attaining better long-term funding performance (Sun et al., 2021). It has also been shown to increasingly lead to practical research that can foster innovation differences in academic institutions (Brown et al., 2015; Dahm et al., 2021).

Designers have working traits consistent with interdisciplinary researchers. According to Brown, Deletic and Wong (2015), interdisciplinary research requires T-shaped research practice and practitioners. This is also a notable trait of the design profession and design practitioners. Kelley and Littman (2006) describe T-shaped individuals as excellent cross-pollinators offering innovative ideas across a team.

12.7   DESIGN AND INTERDISCIPLINARY COMMUNICATION

Ren (2019) states that the main challenges to conducting interdisciplinary research are communication and finding languages to understand each other's discipline research methods. In addition, Wear (1999) contends that the challenge to interdisciplinary discourse is "learning the language" of each disciplinary domain (p. 299). Similarly, Brown et al. highlight that an environment of 'constructive dialogue' needs to be created to empower researchers across disciplines. They cite that focusing on communication can foster empathy and respect for different disciplinary norms and can bridge different research and methodologies and communication cultures (Brown et al., 2015). In this regard, Ren (2019) advises that researchers need to have conversations to identify common ground. Design methods have been proven to facilitate the growth of interdisciplinary research and culture (White et al., 2021) and seem to be particularly effective in facilitating communication between disciplines (White and Deevy, 2020).

12.8   SUMMARY POINTS

The following are summary points from the literature:

1. The development in understanding design as a thought process for addressing complex problems has evolved extensively over the past century.
2. DT emerged as an interdisciplinary composite in the 1990s and 2000s. It quickly grew as a problem-solving methodology for business innovation and extended its range to be applied to address diverse complex problems.
3. DT has been criticized for its reach and short-term focus.
4. DT is now at a transitional stage, there is still a lack of understanding of whether, why, and when it contributes to innovation.
5. To understand its future, there is a real need to reflect and report on its long-term value and use.
6. DT in long-term projects has rarely been reported and or reflected on, and when it has it usually has been focused on implementation.
7. DT is an interdisciplinary composite that is continually evolving and as a process is required to be inclusive across disciplinary domains.
8. Designers have working traits consistent with interdisciplinary researchers.
9. The main challenges to conducting interdisciplinary research are in communication and finding languages to understand other disciplines.
10. Design methods have been proven to facilitate the growth of interdisciplinary research and culture.

12.9   WHAT WE INTEND TO UNDERSTAND

From the ten summary points outlined, we identified a gap in understanding the long-term value of DT and that there were benefits to universities and educational institutes using DT for interdisciplinary research.

In this chapter, we present a long-term case study of using DT to develop interdisciplinary research for aging adults at a university. Through this long-term understanding, we seek to address the question: Does DT work in the long term? Through this question we aim to understand the conditions of embedding long-term DT and offer recommendations for future researchers and practitioners to optimize it in long-term projects.

12.10   THE CONTEXT

The world's aging population is growing at an ever-increasing rate, and this presents complex research challenges for the years ahead. In 2020, the global population over 60 years old reached more than 1 billion people and this is projected to reach nearly 2.1 billion by 2050. Worryingly, at least 142 million older persons worldwide are unable to meet their basic health needs (World Health Organization, 2020).

Recognizing the complex challenges this global demographic shift presents, in 2016, McMaster University, a research-intensive university in Hamilton, Canada, and data collection site for the Canadian Longitudinal Study on Aging, sought to develop an interdisciplinary and impact-driven research institute focused on aging. This is when McMaster Institute for Research on Aging (MIRA) was created, with its objective to optimize the health and longevity of the aging population through leading-edge research, education, and stakeholder collaborations (MIRA, 2024). McMaster's goal in founding MIRA was to create an interdisciplinary research institute that connects existing researchers, research, platforms, and projects in aging, and drives innovation by identifying research questions that address 'wicked problems in aging'. An example of a wicked problem identified was mobility in aging. Mobility is a multidimensional context that includes not only the physical attributes of the person, but also how they engage with their environment using transportation or social engagement. To understand the underlying problems of mobility requires various disciplinary lenses.

MIRA was tasked with a complex challenge: identifying promising research questions and connecting them to the capacity to address such questions with the platforms, expertise, funding, and data available at McMaster. Furthermore, executing research in ways that would meet the mandate of the new institute – supporting research impact and improving the lives of older adults.

This required an approach that differed from conventional academic research, where discovery to impact can typically take an average of 17 years (Morris et al., 2011). With this in mind, MIRA integrated DT into its mandate. The goals of employing it were:

1. To increase interdisciplinary collaboration among aging researchers in McMaster's six faculties.
2. To ask research questions that are truly interdisciplinary in nature and cannot be solved without equal representation of wide variety of perspectives and disciplines.
3. To facilitate user and stakeholder engagement in research.

MIRA is now in its 7th year using DT (2017–2024) and has developed in scale and range of offerings. MIRA members involve researchers across 46 departments and all 6 faculties at McMaster University, including Business, Natural Sciences, Health Sciences, Humanities, Social Sciences, and Engineering. It is made up of over 180 researchers and over 100 trainees across multiple disciplines and career stages. Using a case study approach, we describe DT's role in this development. From this, we seek to address the question: Does DT work in the long term? If so, or not, what are the conditions for embedding long-term DT, and how can it be optimized in long-term projects?

12.11 METHODOLOGY, CASE STUDY APPROACH AND ANALYSIS

Conducting, reporting, and documenting this study posed a complex challenge due to several factors, namely (1) it involved multiple events and activities, (2) it occurred over an extended period (7 years), (3) it involved multiple sources of data and evidence. A case study approach was chosen as it allows for sense-making of disparate sources of data and the reporting of multiple events in a real-life context (Crowe et al., 2011). Reinforcing this decision, Ebneyamini and Sadeghi Moghadam state that "a case study approach is most suited for in-depth research using a holistic lens across multiple events and sources of data" (2018, p. 4). According to Creswell, case studies can explore an in-depth program, event, activity, process, or one or more individuals, where bounded by time and activity, researchers collect detailed information using a variety of data collection procedures over a sustained period (Creswell, 2014; Priya, 2020).

In deciding on a research design for a case study, a comprehensive 'catalogue' of research designs does not exist, therefore this requires researchers to create an appropriate case study strategy for their specific case (Ebneyamini and Sadeghi Moghadam, 2018). Yin (2018) suggests a structured four-stages approach of: (1) designing the case study, (2) conducting the case study, (3) analyzing the case study evidence, and (4) developing the conclusions, recommendations, and implications.

The design of the case study involved cataloguing multiple sources of data relating to using DT at MIRA across seven years, together with the experiences of researchers embedding it. The objective was to ascertain if DT worked over the seven years at MIRA and, if so, seek to understand the conditions of embedding long-term DT.

The case study was conducted and analyzed by two researchers, one researcher was an external experienced design academic and practitioner, and the other was a research manager at MIRA. Conducting and analyzing the case study involved first compiling all the sources of data and all occurrences of the use of DT across the seven years, into chronological order. This was compiled in a Word document and coded into themes of (1) dissemination activities and (2) facilitation and capacity building activities. These themes with descriptions can be seen in Table 12.1. A narrative account of occurrences with reflections was then written in full. In developing conclusions to address the research question: does DT work in the long term? and if so, what were the conditions for embedding it? The narrative account was read and re-read to interpret meaning then reviewed, edited, and 'data reduced' by manually coding into themes (Miles and Huberman, 1994).

In conducting the case study, the epistemological approach from the researchers was both critical and interpretive. Being critical involved questioning one's assumptions of the data in the case study, considering the wider political and social environment of the case. Being interpretative involved understanding meanings/contexts and processes in the case study as perceived from different perspectives, and continually attempting to understand individual and shared meanings to build theory. Adapted from (Crowe *et al.*, 2011).

TABLE 12.1

Source Data Relating to the Case Study

Dissemination Activities

Description	*Quantity*
Journal publications (peer reviewed)	5
Conference publications and presentations (peer reviewed)	8
Invited talks (non-peer reviewed)	4

Facilitation and Capacity Building Activities

Description	*Quantity*
Design thinking workshops and seminars with researchers	12
Design thinking short exercises and design sprints with researchers	10
Design thinking and co-design workshops with older adults and other stakeholders[a]	5
Design thinking workshops with external partners	1
Funding schemes for and including design thinking	187[b]
Feedback sources from members[c]	3
Design thinking resource and repository creation	1

[a] Recorded here are workshops directly facilitated by MIRA. Research teams in turn went on to conduct design thinking workshops of their own, which are not recorded.

[b] This number is made up of 120 student/trainee projects, 62 small catalyst grants and 5 major programs of research.

[c] These include (1) an anonymous member-wide survey, (2) one-on-one interview with Co-PIs x8, and (3) annual reports (from 2018 to 2023).

Reliability and validity in this case study were obtained by an 'auditing approach'. Auditing involves ensuring that records at various stages of the research are maintained and that a peer acts as an auditor to analyze the dependability of the research (Guba and Lincoln, 1994; Priya, 2020). We used a means of peer debriefing with two external 'auditors'. Peer debriefing is a process of exposing research to a peer external to the study, to explore aspects of the enquiry that might otherwise remain subjective (Guba and Lincoln, 1994). We also ensure validity through a process of triangulation, which is a procedure where "researchers search for convergence among multiple and different sources of information to form themes or categories in a study" (Creswell and Miller, 2000, p. 126).

12.12   LIMITATIONS

There are certain limitations in this study. Firstly, with a case study approach, there are challenges to gaining specific impact measurements across multiple uses of DT methods and projects. Though we have started this process in the past 2 years, this will require future research and analysis, and further studies will be required to specifically measure impact across time. Secondly, this case study is specifically using DT in an educational environment rather than that of a business/corporate environment. Thirdly, case studies, due to their nature of qualitatively assessing complex data, can in some cases lead to researcher bias and may give the researcher a position that is too privileged (Crowe *et al.*, 2011). We offset the risk of this by including a peer debriefing process to negate any subjectivity or bias as outlined above.

12.13   CASE STUDY DESCRIPTION

To describe the case study, we outline a summarized narrative account of occurrences across the seven years. We divide the use of DT into a timeline of three phases as follows:

1. Launch and grow phase (2017–2020)
2. Challenges and opportunities (2020–2022)
3. Maturity phase: Understanding impact and outcomes (2022–present) (Figure 12.1)

As detailed in Table 12.1, a wide variety of DT activities occurred across these phases. A holistic and iterative approach was taken to DT from the outset and across the 7 years.

Through consultation with several design practitioners and academics, many avenues were explored to commence and implement DT. To commence and progress activities, different frameworks were considered interchangeably depending on the project's need. These included but were not limited to:

- Stanford D School/Hasso Plattner 5 Stages of Design Thinking: Empathize, Define, Ideate, Prototype and Test (Hasso Plattner, 2013)
- The Double Diamond Model (Design Council, 2019)
- Desirability, Feasibility, and Viability (Brown, 2009)
- Opportunity, Need, and Possibility (Fleisig et al., 2011)

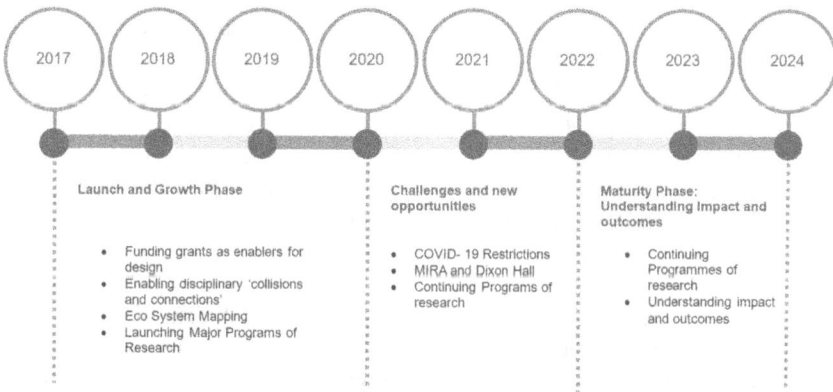

FIGURE 12.1   Timeline of phases where design thinking was utilized across seven years at MIRA.

However, relying solely on standardized frameworks proved difficult, and ultimately, each activity required bespoke design depending on need. Each activity also required deep consideration of the best use of people's time, and therefore each activity required tailored schedules and formats.

It is important to note that DT activities are still ongoing and continuing at MIRA. The purpose of this chapter is to understand a seven-year use and evolution to shape its use in the years ahead. A timeline of phases where DT was utilized across seven years at MIRA is outlined in Figure 12.1

12.14   2017–2020 LAUNCH AND GROWTH PHASE

MIRA launched in 2017 with 35 researcher members from across all six of McMaster's Faculties, and a philanthropic gift of $15M to administer in support of interdisciplinary, impactful research on aging. An early goal was to further identify researchers at McMaster with expertise and interests in aging. Many researchers within the health and social sciences had obvious connections to aging research, but MIRA sought to connect with those in disciplines such as engineering, business, economics, and communications, whose expertise could be applied to aging problems. The first years of MIRA were thus characterized by growth, enabled by targeted, interdisciplinary funding calls, opportunities for "collisions" and connections, and structured approaches to building strategic programs of research.

12.15   FUNDING GRANTS AS ENABLERS FOR DESIGN

One of MIRA's early tactics was to leverage interdisciplinary approaches to foster empathy, ideation, and iteration in research, the building blocks of DT. MIRA's first round of research funding, small 'catalyst or seed' grants, had specific requirements: (1) teams had to include at least three researchers from three *different* Faculties (for example – Health Sciences, Business, and Engineering); (2) they had to catalyze a larger goal, project, or collaboration by collecting exploratory or pilot data, building a platform, and/or generating relationships among researchers and stakeholders, and (3) they must strive to engage stakeholder or end users, with ultimate goal of positively impacting the lives of older people. Researchers responded to these grants with mixed enthusiasm, while some were pleased to see a lever that would support collaboration and connections with stakeholders, others objected with feelings their existing approaches were not interdisciplinary enough, or that there were truly no useful opportunities to engage with stakeholders or end users in their research. This funding mechanism helped to drive the early growth in MIRA membership. Project leads and team members were required to become members as a condition of funding, and outreach and promotion of funding calls resulted in membership growth among other researchers as well. By the end of 2018, MIRA membership had grown from 35 to 98 members (MIRA, 2019).

12.16   ENABLING DISCIPLINARY 'COLLISIONS AND CONNECTIONS'

In addition to funding selected projects, MIRA also dedicated resources to enabling "organic" or casual connections by hosting seminars on aging, meetings and events aimed at an interdisciplinary audience, connecting researchers with potential collaborators beyond the networks and boundaries encountered at departmental seminars. MIRA researchers have reported novel projects and collaborations as a result of attending these events, particularly the MIRA "One Topic, Two Disciplines" series, where two researchers came together to present disparate approaches to problems in aging such as: dementia and driving; polypharmacy and deprescribing intergenerational trauma and life course; and aging and the digital divide in technology access.

As a funder, MIRA coordinates the review and ranking of internal grant and scholarship applications, involving members of its network to assist in the review process. Through the course of seven years, faculty and senior trainee reviewers have applied MIRA evaluation rubrics to the review of interdisciplinary proposals, providing increasingly insightful feedback and suggestions with respect to incorporating DT, stakeholder engagement and interdisciplinarity. Participation in the review of funding proposals has increased researchers' buy-in, fluency and creativity in incorporating design into their research.

12.17   ECOSYSTEM MAPPING

As MIRA's interdisciplinary connections grew, a more structured understanding of these connections was required. An 'Ecosystem Map' (Figures 12.2, 12.3, and 12.4) was designed to accommodate this. The Ecosystem map consisted of a simple 2x2 Matrix, with spectrums of Policy and Application to Theory and Discovery (horizontal) and Policy to Product or Service (Vertical). The boundaries of the quadrants of the matrix were decided by MIRA members to assist frame all researcher's activity at MIRA. The purpose of this map was to visually represent the depth of research expertise on aging within MIRA, to allow MIRA researchers to view themselves as part of a broader community, and to facilitate connections by highlighting areas of strength and opportunity among MIRA members. In a workshop setting, MIRA faculty researchers were asked to place themselves on this map following the instructions:

1. Thinking about your research in aging, place yourself (with a dot) on the figure. While most researchers' work is multi-faceted, for this exercise, please think of your work as it might be viewed by collaborators or grant evaluators.
2. Where is your expertise most concentrated? Horizontal axis: What is the main subject of your research? Policy and Application or more towards Theory and Discovery? Vertical axis: Please consider your research output (or desired output). If that output is solely academic, your dot should be close to the center of the graph. If the outcomes of your work tend toward policy or products and services, you should place yourself closer to the top or bottom, respectively.

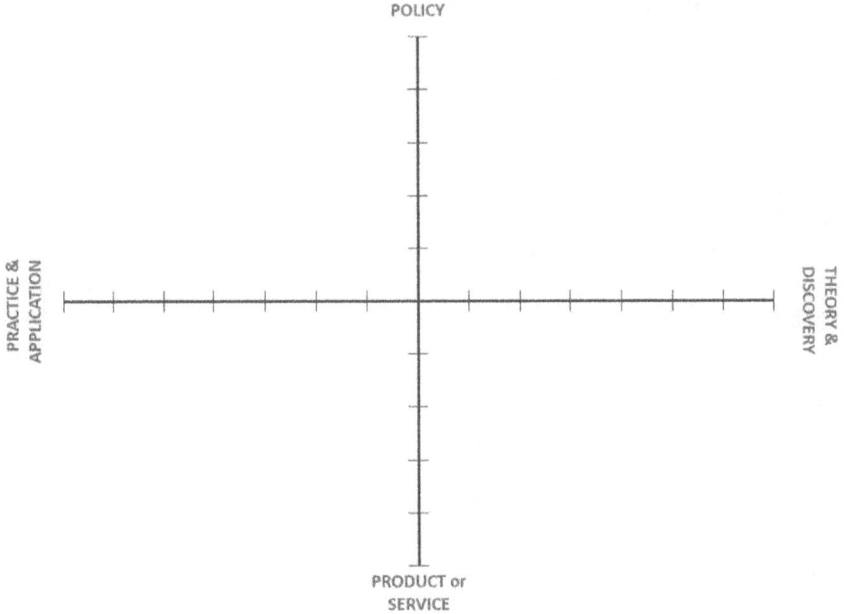

FIGURE 12.2 The ecosystem map consisted of a 2×2 matrix, with spectrums of policy and application to theory and discovery (horizontal) policy to product or service (vertical).

The Ecosystem mapping allowed MIRA members to understand relationships with other researchers and to gain a deeper understanding of other discipline's research on aging. This also allowed for interdisciplinary collaboration in future projects. For example, if a specialist gap was identified in a team or project, the map would display a potential solution to fill that gap. Following a

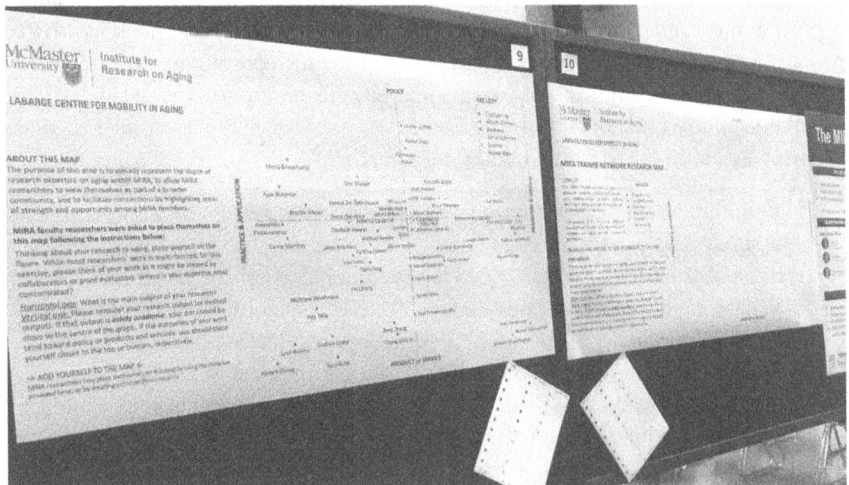

FIGURE 12.3 Ecosystem map in workshop setting.

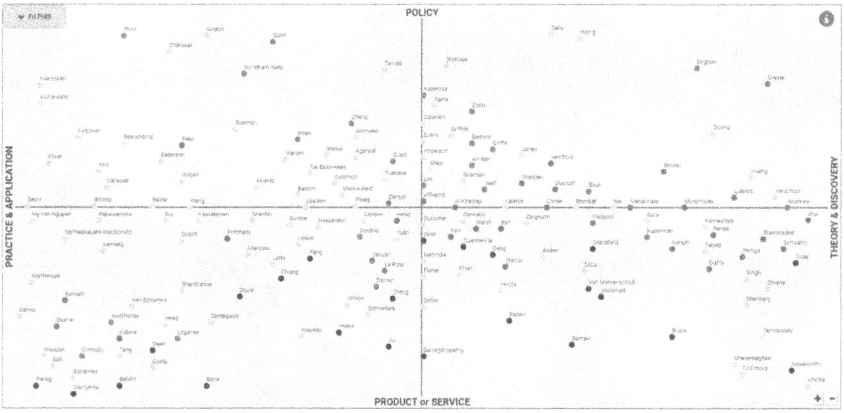

FIGURE 12.4 The ecosystem map was developed into an online platform for MIRA members.

successful workshop with the (paper) map (Figure 12.3) an online digital version of the map was created with dropdown filters for members, faculty, projects, and grants (Figure 12.4).

12.18 LAUNCHING MAJOR PROGRAMS OF RESEARCH

In addition to funding smaller grants and trainee scholarships and fellowships, MIRA funds larger, directed projects, for example, 'Major Programs of Research' (MPRs). These projects were developed using a high-touch approach; MIRA would identify a priority challenge area in aging and invite researchers with interest and expertise to come together and co-develop proposals.

Commencing in 2018, the process of building these projects occurred over approximately 12 months, and employed the following steps:

1. Firstly, through scoping meetings, where MIRA outlined the research topic or challenge to be addressed, and the requirements for a successful proposal. The researchers then shared their relative expertise, platforms, tools, and projects relevant to the topic.
2. MIRA facilitated DT workshops with researchers to elicit potential areas for exploration that are aligned with the opportunity (in this case $1M to fund a project over four years), possibility (what can be done with the expertise, equipment, and other research resources), and need (to be determined in conjunction with stakeholders). These DT workshops sought to begin communication across disciplines, firstly communicating a plain English version of their research and then drawing group conversations about their research further to engage participants in framing potential research proposals in aging. LEGO Serious play was used to promote non-verbal play and to explore interdisciplinary dynamics (White et al., 2023; Figure 12.5).

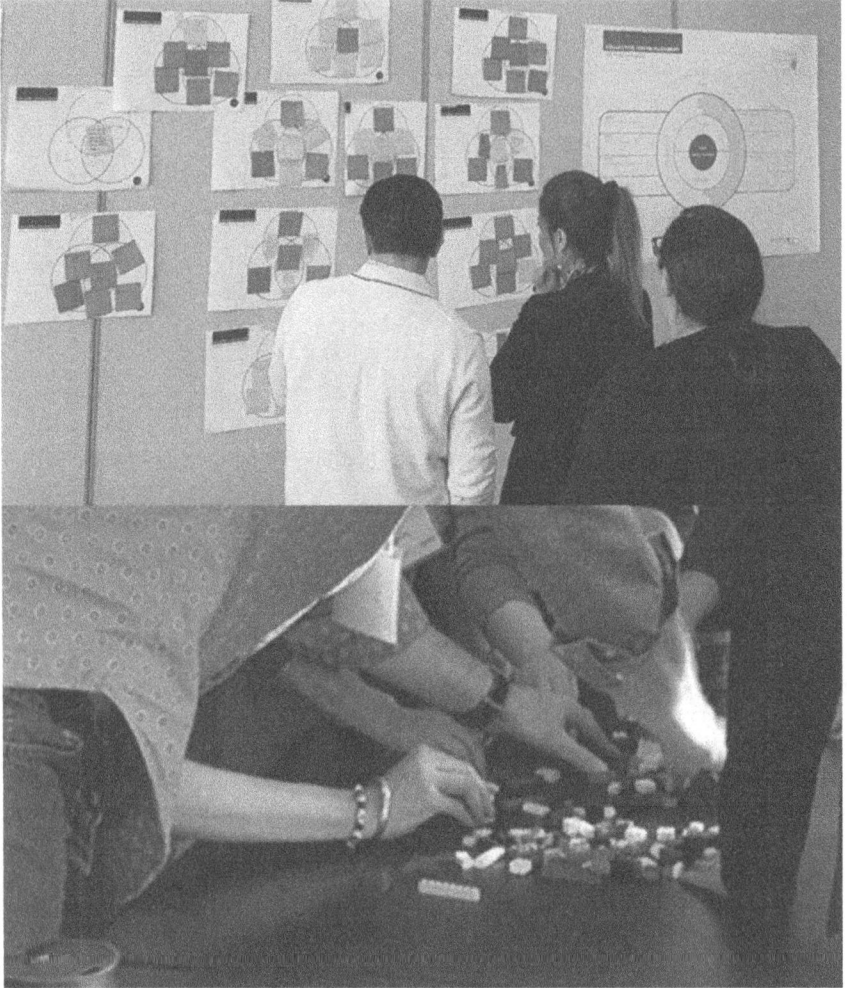

FIGURE 12.5   Workshop activities designed specifically to facilitate IDR connections: DT workshops sought to begin communication across disciplines and engaging participants in framing potential research proposals in aging. This included LEGO serious play to promote non-verbal play and to explore interdisciplinary dynamics (White et al., 2023)

3. MIRA then facilitated opportunities to connect with stakeholders, including older adults, caregivers, health care and other service providers, policymakers, industry partners etc., to iterate on research questions, approaches, and the ultimate goal of creating interdisciplinary projects.
4. Researchers drafted a letter of intent for submission to MIRA and its International Scientific Advisory Committee (ISAC) for review and feedback.
5. Research teams received and integrated feedback, including suggestions for additional disciplines and stakeholders to be involved.

6. Teams submitted a full proposal to MIRA and its ISAC, again subject to iteration as recommended by this review. Once the final proposal was approved and funds were released, MIRA continued to stay engaged with project teams, offering support, suggestions for further funding, and connecting to research partners that may be a good fit for implementation and scale-up.

The first two MPRs funded through this process – EMBOLDEN and MacM3 were launched in 2019 and have built sustainable platforms and attracted large external grants to extend their reach and scope. An assessment of progress (semi-structured interviews with researchers) after 12 months of conducting the MPR Design Thinking workshops found the workshops beneficial in finding 'common languages' in group communication (White et al., 2019).

12.19   2020–2021 CHALLENGES AND NEW OPPORTUNITIES

In early 2020, as the world grappled with COVID-19 and ensuing lockdowns, aging-related research took on new importance and areas of focus. With vaccination and immunity in older priority populations, outbreaks in long-term care, social isolation, and transitioning of services (and really everything possible) online, all these things impacted older adults in profound ways. COVID restrictions highlighted that the needs in aging required agile and flexible approaches to adapt (White et al., 2020; Marston et al., 2020). While continuing with the MPR's, MIRA expanded its funding and other mechanisms of support to explore aging and COVID-19, using DT as its approach.

In 2021, MIRA was presented the opportunity to partner with Dixon Hall, a large social services organization serving a diverse population in Downtown Toronto, with the goal of bringing together lived experience and on-the-ground expertise with evidence-based interventions and evaluation.

12.20   COVID-19 RESTRICTIONS (2020-2021)

Throughout this period, MIRA was well-supported by its culture of DT and collaboration. For example, early in 2020, as researchers were sent home from the University and nearly all research activities halted, MIRA coordinated a series of co-design workshops in the form of 'Idea Exchanges', facilitating brainstorming of actions that would allow researchers to quickly adapt to the challenges presented by lockdowns, and ideating new, timely research questions (White et al., 2021). Researchers shared strategies, such as readying a 'go-bag' that would allow essential research staff to bring previously 'immobile' work home with them in the event of quarantine lab-shutdown, and resources, such as data repositories and calls for papers that would enable research productivity to continue. Researchers also shared ideas to increase the capacity to do research remotely by creating virtual collaboration spaces to rethinking stakeholder engagement (White et al., 2021).

MIRA researchers quickly leveraged an existing longitudinal study – the Canadian Longitudinal Study on Aging generating early evidence that mild COVID infection led to lasting declines in functional mobility in older adults one year after infection (Beauchamp *et al.*, 2022).

12.21   MIRA AND DIXON HALL (2021)

The launch of the MIRA | Dixon Hall Centre in 2021 occurred while much of Hamilton and Toronto was still in lockdown and was precipitated by a slightly rushed introduction between MIRA and Dixon Hall – a social service agency supporting marginalized and at-risk older adults. The impetus to collect COVID-19 data quickly seemed to present a perfect opportunity for the still-forming collaboration to work together. However, the administration of lengthy questionnaires left clients and staff with a view of research participation that felt one-sided and without direct benefit to users. As lockdowns lifted, and Dixon Hall services began to return to normal, MIRA | Dixon Hall Centre's newly appointed Associate Scientific Director set about establishing trust and rapport, through DT, and framed priority-setting exercises.

MIRA facilitated a town hall with over 50 Dixon Hall front-line staff and members of the leadership team ('users'), in brainstorming pain points for older adults and service providers, visualizing ideal future states, and identifying gaps in knowledge and practice (Figure 12.6). The output of that town hall was collated and organized into eight identified priority areas, which the users then ranked and expanded on. These priority areas formed the basis for research funding calls administered by the MIRA | Dixon Hall Centre and are guiding the implementation of new research at the Centre. Additional actions resulting from this process include a "Snacks & Science" knowledge translation series, where McMaster researchers are invited to Dixon Hall to share the latest evidence and relevant information on the priority topic areas in their respective fields of expertise, and

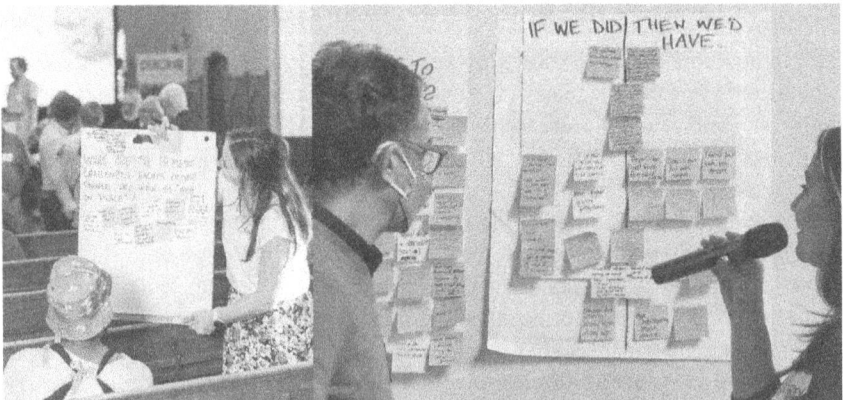

FIGURE 12.6   DT workshops in Dixon Hall.

the formation of a Dixon Hall advisory committee that is consulted for new and proposed research projects. This has been a particularly successful application of DT and is likely attributable to the high degree of buy-in at all lesvels, and the follow-through in applying specific, observable actions.

12.22 2022–PRESENT: UNDERSTANDING IMPACT AND OUTCOMES

With work ongoing, since 2017, MIRA has funded 187 projects, including 67 research grants and 120 student or trainee awards and fellowships, using DT approaches throughout. Understanding impact and outcomes is now ongoing. Open projects report to MIRA on progress, academic output, and other areas of impact each year. In 2022, for example, 32 research grants and 44 trainee projects were open and reported to MIRA on metrics attributable to MIRA support, such as increasing time allocated toward aging research, having received additional funding to support the project, growth in the network of aging-related stakehold-ers, increased collaboration with other disciplines, increased media exposure for their research, as well as an academic publications or conference proceedings.

In this sample, most researchers (69%) and trainees (72%) reported that MIRA support increased the amount of time allocated toward aging research (Patocs et al., 2022, 2023). A third of both researchers and trainees received additional, external funding to support their projects. Further, researchers and trainees described meaningful changes to their research experiences and relationships attributable to their new collaborations (Patocs et al., 2022). For example:

> *"Collaboration with the McMaster Institute for Transportation and Logistics meant a new side project analyzing older adults' responses to a travel survey."*

Postdoctoral fellow, School of Earth, Environment & Society

> *"Input from the new connections has made the methodology and the goals more realistic and yet more meaningful."*

Postdoctoral fellow, Department of Surgery

In an attempt to understand how MIRA's implementation of design in the cul-ture of research may have impacted academic impact, we analyzed the number of peer-reviewed publications trainees recorded against their reported new collabo-rations as a result of MIRA support, within and outside of their discipline, as well as with external stakeholders. Figure 12.7 shows how publications increased lin-early with increasing numbers of academic collaborators suggesting a benefit to both the trainee's career and the potential impact of their research. Interestingly, the number of publications increased for trainees with a few stakeholder collabo-rators relative to those who have zero, but there appeared to be a cost to having a larger number of stakeholder collaborators for trainees reporting four or more.

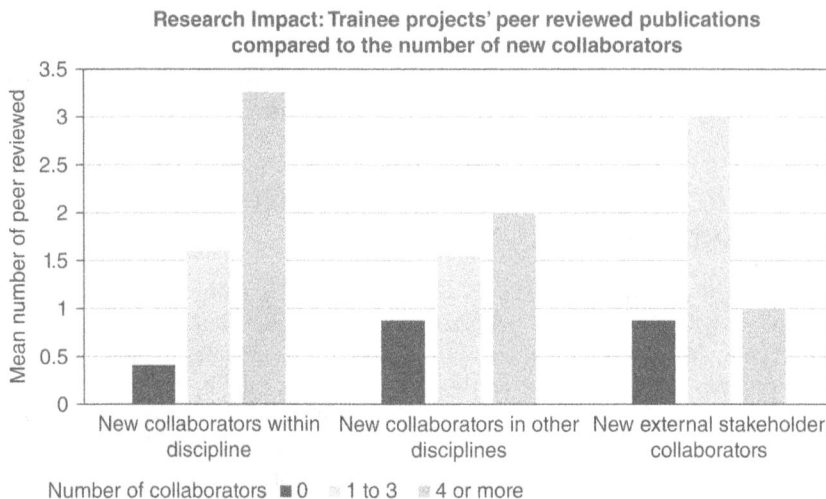

Research Impact: Trainee projects' peer reviewed publications compared to the number of new collaborators

Number of collaborators ■ 0 ▨ 1 to 3 ▨ 4 or more

FIGURE 12.7 Shows how publications increased linearly with increasing numbers of academic collaborators. (Patocs et al., 2023)

This cost may be explained by the amount of time associated with building relationships with external stakeholders. These data do not allow us to determine if this pattern persisted beyond the period for which the grants were considered open by MIRA; it is possible that, with time, this cost of increased stakeholder engagement would disappear or reverse direction entirely.

12.23   10 SUMMARY POINTS FROM THE CASE STUDY

The following are summary points from the case study:

1. The first two MPRs funded through this process have built sustainable platforms and attracted large external grants to extend their reach and scope.
2. *Catalyst grants*: Researchers initially responded with mixed enthusiasm, while some were pleased to see a lever that would support collaboration and connections with stakeholders, others objected with feelings their existing approaches were not interdisciplinary enough, or that there were truly no useful opportunities to engage with stakeholders or end users in their research.
3. Researchers found that DT workshops were beneficial in finding 'common languages' in interdisciplinary group communication.
4. Ecosystem mapping helped visualize an interdisciplinary community of researchers, projects, grants, and other activities.
5. Participation in the review of funding proposals has increased researchers' buy-in, fluency and creativity regarding incorporating design into their own research.

6. DT helped pivot researchers through the unprecedented change of COVID-19 restrictions.
7. Successful application of DT is attributable to the high degree of buy-in at all levels (Dixon Hall).
8. Most researchers (69%) and trainees (72%) reported that MIRA support increased the amount of time allocated toward aging research.
9. A third of both researchers and trainees received additional, external funding to support their projects. Further, researchers and trainees described meaningful changes to their research experiences and relationships attributable to their new collaborations.
10. Trainees who engaged in greater interdisciplinary and stakeholder engagement had greater research impact, as measured by traditional academic metrics.

12.24 LONG-TERM DESIGN THINKING: DID IT WORK?

In conclusion to the case study, and to address the question: Does DT work in the long term?, we can deduce that DT has assisted in facilitating and initiating interdisciplinary research in the long term, therefore it is working. However, the process of DT is not without its challenges, the shortcomings outlined earlier in this chapter such as reach and short-term focus do at times limit the approach. Therefore, for DT to be sustainable in the long-term, certain specific conditions of embedding it need to be in place. These are (1) leadership and buy-in, (2) flexibility and adaptability, and (3) time and support (Figure 12.8).

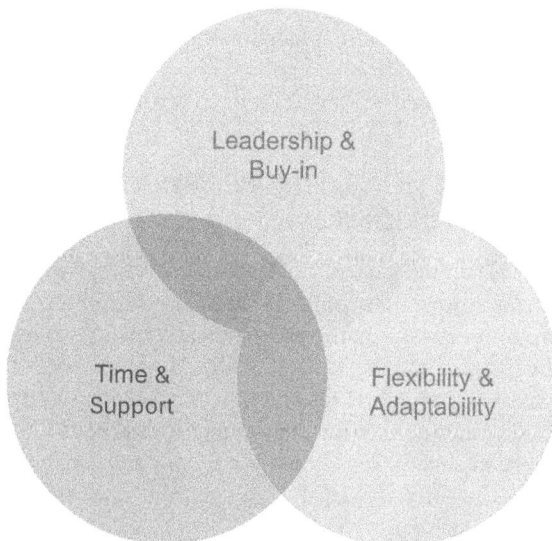

FIGURE 12.8 Conditions of embedding long-term design thinking.

12.25   LEADERSHIP AND BUY-IN

The DT process requires buy-in and understanding from multiple stakeholder perspectives, namely, those who are implementing it and those adopting it. Leadership and champions of the process matter to its long-term success.

12.26   ADAPTABILITY AND FLEXIBILITY

DT is not a one-size-fits-all approach but it is a toolkit of methods and approaches depending on requirements. In this regard, DT frameworks in isolation are, in some cases, not sufficient and require adaptation and flexibility depending on the needs of the project and team. A deep understanding of context is required to adapt and be flexible. Tailoring DT to the specific need offers both the optimum use of time and respect for the use of people's time, increasing participant buy-in. It is recommended for its future development, DT requires continual inspiration from other disciplines to evolve and adapt e.g., anthropology, and sociology, and analogous approaches e.g., co-design and human-centered design.

12.27   TIME AND SUPPORT

An obvious critical component of long-term projects is time itself, and the use of it.

DT requires time to culturally embed over a sustained period, and support needs to be provided for it. A "one and done" short-term focus of DT does not work; (consistent with the literature review at the start of the chapter) this includes one-off high-level workshops. To culturally embed DT, a long-term strategy is required, included in this is time to understand participant's needs in DT and to create positive supportive environments for it to grow. Time to build leadership and buy-in and time to allow for adaptability and flexibility.

12.28   FINAL REFLECTIONS, FUTURE RESEARCH,
AND CONCLUSIONS

The application of DT in the context of MIRA's leadership, funding and facilitation of aging research was, and continues to be a complex undertaking. DT is implemented at several levels – at the project level (of which there are over 100), at the level of the Institute and its practices and policies, and through training and experiences focused on individual researchers, trainees, and leaders. This broad approach has resulted in a culture that has largely embraced DT, and one that has identified and elevated champions whose own actions and capabilities are driving DT and many of its positive results. For this reason, establishing causation for positive impact is challenging.

A further challenge is that true impact still takes a long time. Several MIRA-funded projects have been instrumental in establishing policy: for example, a MIRA-funded postdoctoral fellowship contributed critical immunological vaccine response data that informed the prioritizing of a third COVID-19 vaccine among long-term care residents. During the same period, a MIRA researcher (whose funded project identified linguistic markers of social isolation) was called upon to testify to the Canadian federal government about the impact of social isolation on older adults. These are excellent examples of research mobilization, but the speed at which they were taken up is uncommon and can be partly attributed to the COVID-19 crisis and the unprecedented multi-sectoral focus on evidence synthesis and generating solutions.

Many of MIRA's larger DT-facilitated projects are still underway, and their full impact remains to be seen. At the Institute level and among MIRA members, the impact of DT is apparent. Newly recruited researchers and trainees are often quickly connected to MIRA, rapidly expanding their network of interdisciplinary connections and stakeholders in the aging research landscape. For future research, we will continue to measure impact to obtain longer-term quantifiable results. For now, we can conclude that long-term DT is working and has assisted in facilitating and initiating interdisciplinary research. We can also identify long-term conditions for embedding DT as being: (1) leadership and buy-in, (2) flexibility and adaptability, and (3) time and support.

REFERENCES

Ackermann, R. (2023) *Design Thinking Was Supposed to Fix the World. Where Did It Go Wrong. MIT Technology Review.*

Beauchamp, M. K., Joshi, D., McMillan, J., Erbas Oz, U., Griffith, L. E., Basta, N. E., Kirkland, S., Wolfson, C. and Raina, P. (2022) 'Assessment of functional mobility after COVID-19 in adults aged 50 years or older in the Canadian longitudinal study on Aging', *JAMA Netw Open*, 5(1), p. e2146168.

Blomberg, J., Giacomi, J., Mosher, A. and Swenton-Wall, P. (2017) Ethnographic field methods and their relation to design. In *Participatory Design* (pp. 123–155). CRC Press.

Bousbaci, R. (2008) '"Models of man" in design thinking: The "Bounded rationality" Episode', *Design Issues*, 24(4), pp. 38–52. https://doi.org/10.1162/desi.2008.24.4.38

Brown, R. R., Deletic, A. and Wong, T. H. (2015) 'Interdisciplinarity: How to catalyse collaboration', *Nature*, 525(7569), pp. 315–7.

Brown, T. (2009) *Change by Design: How Design Thinking Transforms Organizations and Inspires Innovation.* New York: Harper Collins, p. 272.

Cai, Y., Lin, J. and Zhang, R. (2023) 'When and how to implement design thinking in the innovation process: A longitudinal case study', *Technovation*, 126, pp. 102816.

Creswell, J. W. and Miller, D. L. (2000) 'Determining validity in qualitative inquiry', *Theory into Practice*, 39(3), pp. 124–130.

Creswell, W. J. (2014) *Research Design-Qualitative, Quantitative, and Mixed Methods Approaches* (4 edn). London: Sage Publications.

Cross, N. (2023a) *Design Thinking: Understanding How Designers Think and Work* (2nd edn). New York: Bloomsbury Visual Arts.

Cross, N. (2023b) 'Design thinking: What just happened?', *Design Studies*, 86, pp. 101187.

Crowe, S., Cresswell, K., Robertson, A., Huby, G., Avery, A. and Sheikh, A. (2011) 'The case study approach', *BMC Medical Research Methodology,*, 11(1), p. 100.

Dahm, R., Byrne, J. R., Rogers, D. and Wride, M. A. (2021) 'How research institutions can foster innovation', *BioEssays*, 43(9), p. 2100107.

Design Council (2019) *What Is the Framework for Innovation? Design Council's Evolved Double Diamond*: Design Council UK. Available at: https://www.designcouncil. org.uk/news-opinion/what-framework-innovation-design-councils-evolved-double-diamond (Accessed: 12th November 2020).

Dunne, D., Eriksson, T. and Kietzmann, J. (2022) 'Can design thinking succeed in your organization?', *MIT Sloan Management Review*, 64(1), p. 60–67.

Ebneyamini, S. and Sadeghi Moghadam, M. R. (2018) 'Toward developing a framework for conducting case study Research', *International Journal of Qualitative Methods*, 17(1), p. 1609406918817954.

Fleisig, R. V., Mahler, H. and Mahalec, V. (2011) "Engineering design in the creative age", *Proceedings of the Canadian Engineering Education Association (CEEA)*. https:// doi.org/10.24908/pceea.v0i0.3683

Gaynor, L., Dempsey, H. and White, P. J. (2018) 'How Design Thinking Offers Strategic Value to Micro-Enterprises'. *Design as a catalyst for change - DRS International Conference 2018*, Limerick Ireland: Design Research Society, 2974–2986. https:// doi.org/10.21606/drs.2018.434

Guba, E. G. and Lincoln, Y. S. (1994) 'Competing paradigms in qualitative research', *Handbook of Qualitative Research*, 2(163-194), p. 105.

Iskander, N. (2018) 'Design thinking is fundamentally conservative and preserves the status quo', *Harvard Business Review*, 5(09), pp. 2018.

Kelley, T. and Littman, J. (2006) *The Ten Faces of Innovation: IDEO's Strategies for Beating the Devil's Advocate and Driving Creativity Throughout Your Organization*. Crown.

Kennedy, F., White, P. J. and Dempsey, H. (2024) 'Improving the door-to-door customer journey for a national public transport Company', *Sustainability*, 16(20). https://doi. org/10.3390/su16208741

Ku, B. and Lupton, E. (2022) *Health Design Thinking: Creating Products and Services for Better Health*. MIT Press.

Liedtka, J. (2018) 'Why design thinking works', *Harvard Business Review*, 96(5), pp. 72–79.

Marston, H. R., Shore, L. and White, P. J. (2020) 'How does a (smart) age-Friendly eco-system look in a post-pandemic society?', *International Journal of Environmental Research and Public Health*, 17(21). https://doi.org/10.3390/ijerph17218276

Martin, R. L. (2009) *The Design of Business: Why Design Thinking Is the Next Competitive Advantage*. Harvard Business Press.

Miles, M. and Huberman, A. M. (1994) *Qualitative Data Analysis: An Expanded Sourcebook*. 2 edn. Thousand Oaks, CA: Sage Publications.

MIRA. (2019) *Annual Report 2019*. Canada: Labarge Centre for Mobility in Aging Collaborative for Health & Aging, McMaster University.

MIRA. (2024) *McMaster Institute for Research on Aging (MIRA) https://mira.mcmaster. ca/about/* (Accessed: 22/05 2024).

Morris, Z. S., Wooding, S. and Grant, J. (2011) 'The answer is 17 years, what is the question: Understanding time lags in translational research', *J R Soc Med*, 104(12), pp. 510–20.

Nussbaum, B. (2011) 'Design Thinking Is a Failed Experiment. So What's Next?', https://www. fastcodesign.com/1663558/design-thinking-is-a-failed-experiment-so-whats-next.

Patocs, A., Alders, G., White, P. J., Dubé, A. and Raina, P. (2022) 'Aging research and design thinking: How interdisciplinary collaboration and stakeholder engagement can drive trainees' research impact.' *CAG2022, 51st Annual Scientific and Educational Conference*, Regina, Saskatchewan, Canada. https://doi.org/10.13140/ RG.2.2.20336.94720

Patocs, A., Alders, G., White, P. J., Dubé, A., Wauben, I. and Raina, P. (2023) 'Measuring impact within an interdisciplinary aging research landscape'. *CAG2023: Community Engaged Teaching, Research & Practice*, Toronto, Canada. https://virtual.oxfordabstracts.com/#/event/4030/submission/613

Plattner, H. (2013) 'An introduction to design thinking', *Institute of Design at Stanford*, pp. 1–15.

Priya, A. (2020) 'Case study methodology of qualitative research: Key attributes and navigating the conundrums in its Application', *Sociological Bulletin*, 70(1), pp. 94–110.

Rittel, H. W. J. and Webber, M. M. (1973) 'Dilemmas in a general theory of planning', *Policy Sciences*, 4(2), pp. 155–169. https://doi.org/10.1007/BF01405730

Ren, Z. J. (2019) 'The rewards and challenges of interdisciplinary Collaborations', *iScience,* 20, pp. 575–578.

Rylance, R. (2015) 'Grant giving: Global funders to focus on interdisciplinarity', *Nature*, 525(7569), pp. 313–315.

Schön, D. (1983) *The Reflective Practitioner; How Professionals Think in Action*. London: Temple Smith, Basic Books.

Shrivastav, S. and Joshi, R. 'Design Thinking for Long-Term Product Planning'. *Design for Tomorrow—Volume 3*, Singapore, 2021//: Springer Singapore, 387-398.

Simon, H. (1968) *The Sciences of the Artificial* (3 edn). MIT Press.

Stember, M. (1991) 'Advancing the social sciences through the interdisciplinary enterprise', *The Social Science Journal*, 28(1), pp), pp. 1–14.

Sun, Y., Livan, G., Ma, A. and Latora, V. (2021) 'Interdisciplinary researchers attain better long-term funding performance', *Communications Physics*, 4(1), p. 263.

van der Westhuizen, D., Conrad, N., Douglas, T. S. and Mutsvangwa, T. (2020) 'Engaging communities on health innovation: Experiences in implementing design Thinking', *International Quarterly of Community Health Education*, 41(1), pp. 101–114.

Verganti, R., Dell'Era, C. and Swan, K. S. (2021) 'Design thinking: Critical analysis and future evolution', *Journal of Product Innovation Management*, 38(6), pp. 603–622.

Wear, D. N. (1999) *Challenges to Interdisciplinary Discourse: Ecosystems*. New York: Springer-Verlag.

White, P. J., Alders, G., Patocs, A. and Raina, P. (2021) 'COVID-19 and interdisciplinary research: What are the needs of researchers on aging?', *Tuning Journal for Higher Education,* 9(1), pp. 239–263. https://doi.org/10.18543/tjhe-9(1)-2021pp239-263

White, P. J. and Deevy, C. (2020) 'Designing an interdisciplinary research culture in higher education: A case Study', *Interchange,* 51(4), pp. 499–515. https://doi.org/10.1007/s10780-020-09406-0

White, P. J., Deevy, C., Casey, B., Patocs, A. and Raina, P. (2023) 'Co-design for interdisciplinary research communities', *IASDR 2023: Life-Changing Design*, 9–13 October, Milan, Italy. https://doi.org/10.21606/iasdr.2023.226

White, P. J., Harrington, L., Patocs, A. and Raina, P. (2019) 'Using Design to Create Interdisciplinary Research in Ageing', *International Association of Gerontology and Geriatrics European Region Congress 2019 (IAGGER 2019)*, Gothenburg, Sweden. https://designchange.wordpress.com/wp-content/uploads/2021/01/iagger-2019-poster.pdf

White, P. J. and Kennedy, C. (2021) 'Designing a module in entrepreneurship for product design students', *Industry and Higher Education*, 36(2), 217–226. https://doi.org/10.1177/09504222211013742

White, P. J., Marston, H. R., Shore, L. and Turner, R. (2023) 'Learning from COVID-19: Design, age-Friendly technology, hacking and mental Models', *Emerald Open Research*, 1(2). https://doi.org/10.1108/EOR-02-2023-0006

Wolff, B. (2023) 'When Silicon Valley Embraced the Arts: Lessons from PAIR At Xerox PARC'. Available at: https://www.forbes.com/sites/benjaminwolff/2023/10/27/when-silicon-valley-embraced-the-arts-lessons-from-pair-at-xerox-parc/

World Health Organization (2020) *Decade of Healthy Ageing: Baseline Report.* Geneva: World Health Organization, p. 203. https://iris.who.int/handle/10665/338677.

Wrigley, C., Nusem, E. and Straker, K. (2020) 'Implementing design thinking: Understanding organizational Conditions', *California Management Review*, 62(2), pp. 125–143.

Yin, R. K. (2018) *Case Study Research and Applications.* Thousand Oaks, CA: Sage.

Index

For Product Safety Concerns and Information please contact our EU
representative GPSR@taylorandfrancis.com
Taylor & Francis Verlag GmbH, Kaufingerstraße 24, 80331 München, Germany